W9-BRY-900

# MEASURING UP: CHARTING PATHWAYS TO MANUFACTURING EXCELLENCE

**The BUSINESS ONE IRWIN/APICS Series in**

**Production Management**

Supported by the American Production
and Inventory Control Society

OTHER BOOKS PUBLISHED IN THE BUSINESS ONE IRWIN SERIES
IN PRODUCTION MANAGEMENT

*ATTAINING MANUFACTURING EXCELLENCE*    Robert W. Hall
*BILLS OF MATERIALS*    Hal Mather
*PRODUCTION ACTIVITY CONTROL*    Steven A. Melnyk and
Phillip L. Carter
*MANUFACTURING PLANNING AND CONTROL SYSTEMS*,
Second Edition
Thomas E. Vollmann, William Lee Berry, D. Clay Whybark
*THE SPIRIT OF MANUFACTURING EXCELLENCE*    Ernest Huge
*STRATEGIC MANUFACTURING: DYNAMIC NEW DIRECTIONS
FOR THE 1990s*
Patricia E. Moody
*TOTAL QUALITY: AN EXECUTIVE'S GUIDE FOR THE 1990s*
The Ernst & Young Quality Improvement Consulting Group
*TIME-BASED COMPETITION: THE NEXT BATTLEGROUND
IN AMERICAN MANUFACTURING*
Joseph Blackburn

BUSINESS ONE IRWIN/APICS Series in
Production Management

APICS ADVISORY BOARD

L. James Burlingame
Eliyahu M. Goldratt
Robert W. Hall
Ed Heard
Ernest C. Huge
Henry H. Jordan
George W. Plossl
Richard Schonberger
Thomas E. Vollmann
D. Clay Whybark

# MEASURING UP: CHARTING PATHWAYS TO MANUFACTURING EXCELLENCE

*Robert W. Hall*
*H. Thomas Johnson*
*Peter B. B. Turney*

**BUSINESS ONE IRWIN**
Homewood, Illinois 60430

© RICHARD D. IRWIN, INC., 1991

*All rights reserved.* No part of this publication may be
reproduced, stored in a retrieval system, or transmitted,
in any form or by any means, electronic, mechanical,
photocopying, recording, or otherwise, without the prior
written permission of the copyright holder.

This publication is designed to provide accurate and
authoritative information in regard to the subject matter
covered. It is sold with the understanding that neither the
author nor the publisher is engaged in rendering legal, accounting,
or other professional service. If legal advice or other expert
assistance is required, the services of a competent
professional person should be sought.

*From a Declaration of Principles jointly adopted by a Committee
of the American Bar Association and a Committee of Publishers.*

Project editor: Jean Roberts
Production manager: Ann Cassady
Compositor: Publication Services, Inc.
Typeface: 11/13 Century Schoolbook
Printer: Arcata Graphics/Kingsport

### Library of Congress Cataloging-in-Publication Data

Hall, Robert W.
   Measuring up: charting pathways to manufacturing excellence /
Robert W. Hall, H. Thomas Johnson, Peter B. B. Turney.
       p.   cm.—(The Business One Irwin/APICS series in production
management)
   Includes bibliographical references and index.
   ISBN 1-55623-359-0
   1. Manufactures—Management.   2. Production management.
3. Manufacturing processes.   4. Manufactures—Quality control.
I. Johnson, H. Thomas   II. Turney, Peter B. B.
III. Title.   IV. Series.
HD9720.5.H35   1990
658.5—dc 20                                              90–37490
                                                            CIP

*Printed in the United States of America*
1 2 3 4 5 6 7 8 9 0 K 7 6 5 4 3 2 1 0

# PREFACE

Performance measurement is an important component of manufacturing excellence, which represents a departure from industrial practice of a decade ago. When we do something differently we must also measure it differently. Performance measurement is built into the processes of identifying and overcoming problems that lie at the heart of continuous improvement.

Changes in measurement systems lagged behind in the revolution in manufacturing that began in the United States during the past decade. This lag has been a barrier to progress for many companies, and sometimes even has encouraged movement in the wrong direction. The next decade should see a substantial displacement of "old" measurement systems by performance measures that are a part of the revolution.

We, the authors, were fortunate to have been involved in some of the early philosophical thrashing over differences in performance measurement systems. In fact, we thrashed with each other to derive a consistent framework of measurement thought from the diverse experiences of each of us working with different manufacturers. Forcing ourselves to integrate our concepts of measurement has been helpful to each of us. We hope the outcome will stimulate a similar stirring of fresh thought in our readers.

Thanks are due to many people who helped us arrive at this stage in our thinking—too many to mention them all by name. Professor Hall became interested in performance measurement because many readers and contributors to *Target* found it important. The beginning points of much material for this book were conversations with people while editing stories on Motorola, Harley-Davidson, Xerox, and Florida Power & Light Co. There were also some enlightening conversations with consultants at Ernst & Young and Coopers & Lybrand. Over time, the specifics of attribution unfortunately became fuzzy in the mind, but inspiration for much of the material for Chapters 1, 2, 5, and 6 originated from these sources.

Professor Johnson thanks Professor Roy D. Shapiro of the Harvard Graduate School of Business for many invaluable comments and suggestions on earlier drafts of the material in Chapter 3. Johnson also acknowledges that many ideas in Chapter 3 emanated from joint research with members of Arthur Andersen's operations consulting practice, and particularly Robert J. Berling, Jr.

Professor Turney acknowledges the encouragement and influence of Professor Robin Cooper of the Harvard Graduate School of Business Administration in the development of his approach to activity-based costing. He also acknowledges his reliance on Professor Cooper's contributions to activity-based costing theory in writing Chapter 4. Professor Turney thanks the Tektronix Foundation for generously supporting his research activities.

Jim Childs, Senior Editor at Business One Irwin, is really responsible for sparking the idea to write this book. We appreciate his patience while we were struggling through mismatches in ideas and conflicts from busy schedules.

Jan Cashion keyboarded and assembled the final manuscript. In the end, we the authors take responsibility for the correctness and utility of the book.

> **Robert W. Hall**
> **H. Thomas Johnson**
> **Peter B. B. Turney**

Indianapolis, Indiana
April 15, 1990

# TABLE OF CONTENTS

# CHAPTER 1

---

# THE NEW MANUFACTURING
# PARADIGM

---

During the past decade a revolution in thought has been creeping up on manufacturing in the United States and elsewhere. The antecedents of this new type of thinking arose in many places, including the United States, which had such pioneers as Henry Ford and, in recent years, Japan, whose manufacturers have been much noted for their advancement.

With this revolution has come an avalanche of "buzzwords." When manufacturers are being persuaded to change their practices, the terms for new practices take on a powerful appeal. Here are some of the new techniques mentioned in recent books and articles:

- Total quality control
- Total productive maintenance
- Just-in-time manufacturing
- Computer-integrated manufacturing
- Total factory automation
- Total employee involvement
- Quality function deployment
- Performance management
- Design of experiments
- Single-minute exchange of die
- Supply chain management

This is only a partial list of catchphrases. Some are so well-known as to be in common use. Others are rare. But all have something in common: they assume that competitive manufacturing today requires top management to emphasize customer service leading the people of the company to become more quality-intensive, time-intensive, and people-intensive.

Any company pursuing the new manufacturing has companywide improvement programs that cover the three broad areas shown in Figure 1–1: people, process, and quality. In turn, each of these three areas incorporates so many new subconcepts that it is impossible to list them all. The eyes tire and the head swims trying to integrate the different concepts into an overall picture.

**FIGURE 1–1**
**The New Manufacturing Paradigm's Elements**

1. Decentralized responsibility; responsibility to customers
2. Multifunctional: Workers
   Staff
   Management
3. Strong horizontal communication
4. Employee contributions to improvement
5. Great development of employee skills

People

Customer satisfaction

Quality

Process

Total quality management in all functions of the company
1. Strong motivation to serve customers
2. Concept of internal customers
3. Targeting improvement (rather than just conforming to specifications)
4. Techniques
   Statistical process control
   Quality function deployment
   Experimental design
   Routine quality problem solving
   Fail-safe methods

Product design for customer use
   For produceability
   For process quality
Plant
   Setup time reduction
   Flow layouts
   Space reduction
   Total productive maintenance
   *Jidoka*
   Use of cycle time analysis
   for improvement
Systems
   Visible control methods
   Synchronized materials planning
   Performance measurement
Suppliers
   Limited number
   Strong organizational relations
   High quality and short lead
   times from each

Performance measurement is but one part of this overall picture. To measure performance in the new manufacturing, one must first understand what is involved—what is to be measured and why.

The new paradigm cannot be put into a succinct phrase or two. It is a reorientation of common sense from an adjusted point of view, a different perspective about manufacturing. The attempt to catch the spirit of this change technique-by-technique has spawned the births of all the new buzzwords. Here is a quick overview of the new paradigm:

1. Adopt service and operations improvement goals. *Customer satisfaction* is paramount. Improve manufacturing by radically changing how the work is conceived and organized as well as through technology.
2. Quality is a quasi religion.
3. Lead times for *all* activities should be compressed: throughput times, customer response times, product development times, and the like. Value is based on time. Add value by doing something useful to material at all times. Contribute to service improvement at all times. *Time,* rather than capital and economy of scale, becomes a focal point of operations planning, control, and improvement.
4. People in all parts of an organization should be responsible for contributing to customer satisfaction—for the full scope of the business at their location. This implies a radically different role for managers and workers. Serving internal customers weakens top-down control and promotes horizontal communication. It shows a recognition that improvement occurs through people. Develop people, and through them technology can be applied to the problems of the business and the customers.

These points are undoubtedly somewhat cryptic to readers who have not been previously saturated with explanations of the "new manufacturing." It takes time to work through doubts and fears to assimilate the new "excellence" thinking.

Some examples help. The story of Toyota has been the classic, oft-quoted case. It still offers instruction, but examples of other companies provide a more familiar backdrop.[1]

## HARLEY-DAVIDSON MOTOR COMPANY

In the late 1960s Harley-Davidson held almost 100 percent of the North American super heavyweight motorcycle market, defined as machines with engine displacement above 850 cc. Every American knew the distinctive rumble of a Harley at full throttle. Honda, Yamaha, Kawasaki, and Suzuki were just beginning to enter the market with small off-road bikes—junior motocross machines that posed no vital threat. The real performance competition came from European makes such as BMW.

Inside Harley-Davidson, operations were typical of many American, old-line manufacturing companies at the time. Parts shortages were chronic. As much as 70 percent of all bikes finishing assembly were missing at least one part. Bikes sat everywhere awaiting parts. Quality was a problem, too. The majority of finished machines needed rework for one reason or another.

As everyone now knows, the Japanese kept coming. Harley's share of the heavyweight market kept dropping, hitting its low point in 1983 at 23 percent. Both Honda and Kawasaki built plants in the United States so that Harley competed with American-built (or at least American-assembled) heavyweight bikes.

In June 1981 a group of Harley executives bought the company back from AMF in a highly leveraged transaction. There was no money for new technology or new machines, so Harley could not automate its way out or try any form of expensive systems fix. The people at Harley had to change themselves.

After a year or more of hesitation because of the strangeness of the concepts, Harley's managers commenced operational reform in 1982. They began three primary thrusts almost simultaneously and have been building on these bases ever since:

1. Employee involvement (EI) [people]
2. Statistical operator control (SOC) [quality]
3. Material-As-Needed (MAN) [process]

Over time these programs have expanded and overlapped until now Harley-Davidson is organized for continuous improvement.

The employee involvement process began with the simple recognition that managers need to talk with workers, skilled tradespeople, and supervisors, and most of all they need to listen. Harley has developed several areas of employee training under the heading of employee involvement:

- Quality circles
- Preventive maintenance
- Setup time reduction
- Value analysis
- Statistical operator control

As a result the average Harley plant employee is far more trained and capable today than a decade ago, and far more responsible too. Today many of the ideas and improvements come from Harley's frontline people. Most employees were enthusiastic about as well as fearful for the company's future.

Probably the most difficult part of the employee involvement process was getting managers to learn how to function in their new roles: facilitating, evaluating suggestions, providing resource follow-up to implement employee recommendations—very different from their previous role, in which they did all the planning and deciding.

Most of the benefit from preventive maintenance, for example, is that employees who performed basic maintenance on their equipment learned more about it and took more pride in it. For both mechanical reasons and behavioral reasons, machine availability (not machine utilization) gradually increased. Likewise, few ideas to reduce setup times on equipment have come from engineers. The best and least expensive ideas have come from operators, maintenance personnel, and skilled tradespeople who regularly work with the equipment. Statistical operator control is so named because operators can control their process with statistics. Control charts are always kept by hand so that the operator is always personally involved. Some of the ideas to reduce a machine's setup time incorporated tooling revisions to enable it to perform with a narrower process variance as well. *Process capability* is a common term at Harley-Davidson. This means that each process should be capable of defect-free production if properly set up and maintained.

Since 1984 all employees have been expected to complete a 40-hour basics course in statistics. Homework assignments are designed to encourage employees to quickly apply what they learn and to commit themselves to use systematic methods of quality problem diagnosis and resolution.

Harley's Material-As-Needed process is modeled after the Toyota production system. It originated in 1982 primarily to improve quality. An enormous amount of time and energy had been spent reworking parts that had found their way into inventory. The reasoning was that less inventory would create less rework even if the defect rates did not drop. In addition, if less inventory decreased the lag time between operations (whether inside Harley or at suppliers), faster feedback on problems was likely to cause improvement. Experience proved both these assumptions correct.

To cut lead times, Harley revised layouts, eliminated storage spaces, and reduced setup times. Harley also introduced a mixed-model final assembly schedule. Its purpose was to consume an approximately equal amount of all kinds of parts during each day of a one-month schedule period. That made it possible to keep small quantities of many different kinds of parts on hand at the York, Pennsylvania, assembly plant. The schedule for the Milwaukee engine plant, which was derived from the assembly schedule, also called for approximately equal usage of each type of engine throughout each day of the month. Each supplier of parts could also each see a repeating daily schedule.

At the engine plant and in other fabrication areas, the new schedule allowed a regular pattern of setups for different parts. That is, operators could practice setup for the same parts at regular intervals. There were fewer sporadic surprises in requests for parts. Regular work cycles assisted both the reduction of setup times and the resolution of process capability problems.

Inside the plants much material is moved by a *pull system* triggered by *kanban* cards, hence the name, *Material-As-Needed.* At assembly and elsewhere the system allows operators to control the material, which therefore does not pile up. Lead times and waste cannot become excessive if the system is used with discipline to maintain upper limits on the amount of material in pipelines.

There is another benefit. A *Kanban* system requires a fixed routing. Harley-Davidson takes this requirement a step further. Their MAN (*kanban*) cards specify the exact machines they are to travel between. This procedure eliminates the variations caused by machining parts with a random selection of machines (and sometimes tools also).

The Material-As-Needed process differs from the Toyota system in its use of three kinds of cards:

1. *Raw material:* From first operation to raw material stores and back again. Used to move material from raw material to the machine. Also used to trigger reorders of raw material into stores as needed.
2. *Move:* Used to move finished manufactured parts from storage to point of use (usually assembly).
3. *Production:* Used between machines to control parts fabrication via a specific sequence of machines.

The signaling systems established through *kanban* cards can become quite complex. Some companies develop warning lights, multiple colors, special colors for cards added in case of trouble, and so forth. A simplified version of a *Kanban* system (not Harley's) is diagrammed in Figure 1–2, and a few basic rules for its use are given.

**FIGURE 1–2**
**Simplified View of Kanban System**

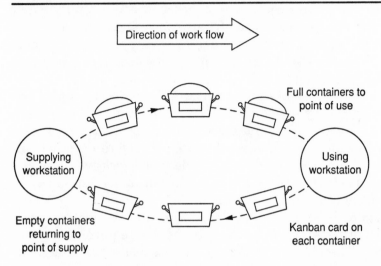

Rules for the *kanban* (or other form of "pull") system:

1. Use only standard containers so that a fixed number of one type of part goes into each container.
2. Fix the number of cards (or containers) for any specific part number between any two workstations (or other points).
3. No card; no production. Do not overbuild.
4. Maintain the number of cards or containers at the minimum level needed for smooth production. That is, lead times between the two work centers should be as short as possible.
5. Production control (signaling for more parts) should be a visible, simple communication between the two work centers, even if the *kanban* signal is sent electronically. It is not a centrally processed control system.

Actually the rules depend upon everyone developing the discipline to use the system. It is not a hard system to understand, but everyone must pay attention, and getting people to appreciate the underlying reasons for the system is a matter of education and practice.

In Harley-Davidson's case, and in others, simplifying the material flow led to simplifying the accounting for materials inside the plants. The amount of material actually in process is kept fixed by the system. The amount of inventory is small, and its throughput time is short. Therefore material does not have to be counted into and out of inventory by transaction. When a motorcycle finishes assembly its bill of material is used to "backflush" the parts from purchased parts inventory. The backflush process explodes the bill of materials through fabrication to relieve parts from raw materials inventory also. In these circumstances, inventory-tracking transactions are time-consuming. They impede the fast flow of material, and they add no information of value to management.

The real use of the system is not for production control or for inventory control. It is to reduce lead times and to stimulate the development of people and processes. A *Kanban* system used *only* for production control is used at only a fraction of its potential power.

Inventory levels are a surrogate measure of lead time. Inventory level divided by production rate equals throughput time (or lead time of material going through a process). Lead time is of most interest and importance. The inventory reduction that occurred at Harley-Davidson is merely a bonus in the form of reducing an unnecessary working capital investment. Reducing lead times demands reform and reorganization of operations.

In Harley-Davidson's case, assembly plant inventory decreased from about 10 weeks' supply in 1982 to about $2\frac{1}{2}$ weeks' supply in 1988. It is also a good idea to think of stating inventory in terms of its time coverage—as a rough indicator of lead times.

Harley-Davidson no longer collects and reports labor hours in detail. Job cards to collect labor have been eliminated, and job classifications combined. Labor and overhead are combined into a single account called conversion cost. About 40 labor control codes that once existed on cost accounting statements no longer exist.

Two other performance measurement changes are notable. First, cost center efficiency data are no longer collected and reported. In the past that measure had driven individuals to overproduce just to lengthen production runs and increase efficiency. With the new thinking, overproduction is a waste that only adds to lead times.

Second, the machine utilization report was abolished. Instead, preventive maintenance efforts are judged in terms of machine availability. Utilization states the percentage of available time operators actually ran a machine. Availability states the percentage of time a machine was capable of performing when it was needed.

In addition, machine capabilities have become important. These state whether equipment is statistically capable of producing parts without defect due to random variations.

While reforming internal operations, Harley-Davidson also cut its total number of suppliers from about 800 to about 400. Many are sole-source suppliers. Cost reduction with suppliers remains very important, but it is generally accomplished by working with an established supplier to improve part design, production processes, or transportation. Switching to a new supplier only occurs when an old one proves unable to improve. Bidding competition is largely a bygone practice.

To bring material in small amounts to the York, Pennsylvania, assembly plant, Harley-Davidson runs two different kinds of freight consolidation methods. Suppliers within a 200-mile radius are served by MANTRAN, a Harley-Davidson truck operated by a Harley employee that completes a daily "milk run" to pick up a mixed load from several different suppliers.

Suppliers at a distance greater than 200 miles ship small lots to consolidation points either in Aurora, Illinois, or Cleveland, Ohio. At these two points shipments from about 60 different suppliers are transferred into two trucks, greatly reducing dock congestion in York. This practice is necessary to keep the cost of shipping small lots in less-than-truckload quantities reasonable. Overall cost is down, and the on-time delivery record from suppliers is "in the high nineties." Few motorcycles today finish assembly minus a part. The suppliers seem to like it, too. According to one, "With Harley, you know you have to put the parts on the truck at the designated time, no ifs or buts, and no matter who must finish the work and load the truck."

By 1987 Harley-Davidson told the federal government that a special tariff enacted in 1983 could be repealed. The company had become competitive with the Japanese manufacturers on its own. By the end of 1989 Harley's share of the super heavyweight motorcycle market was 59 percent and rising. Harley did not claim to be the world's best, but the company had returned to profitability and reissued public stock. The biggest challenge was to sustain the improvement during a period of growth and financial success. All Harley-Davidson personnel from bottom to top must constantly remind themselves that the top goal is customer service, not current financial success.

## TEAM XEROX

Xerox Corporation is the descendent of the people who invented xerography, the copying process usually referred to today as *Xeroxing*. Few stories have so epitomized the commercial success that can follow a technical breakthrough as that of Xerox in the 1960s. Yet by 1980 Xerox recognized that it was in trouble—and against those pesky Japanese. The basic patents had run out, and other companies were proving to be very good at continuously improving the process Xerox had started 20 years earlier. Benchmarks against Japanese competitors told the story. Xerox was way behind in unit cost and quality.

Many of the reforms that began in 1981 concentrated on Xerox's suppliers because 80 percent of the copiers' cost of goods sold was purchased parts, and the competitors' selling prices were about the same as Xerox's factory costs—sometimes as low as the cost of Xerox's purchased material.

Benchmarking other companies' operating performance was an essential factor in motivating Xerox to make drastic changes. It was also the primary "wake-up jolt" administered to Xerox suppliers. Xerox emissaries visited many supplier's top management asking them to restate their company's performance data so as to compare it with the best similar operation Xerox could find in the world.

However, before Xerox could effectively reform its suppliers, it had to take some of the cure itself. That began with total quality control (TQC), which in turn began with courses in statistical process control (SPC). Xerox soon found that full implementation of TQC was impossible without a heavy dose of employee involvement. In order to generate the necessary atmosphere for hundreds of multidisciplinary employee involvement (EI) teams to function simultaneously, the Xerox top command found itself faced with the need for a broad, sweeping change in corporate culture. This change began with CEO David Kearns himself. Known as *leadership through quality*, the new cultural style has three aspects:

1. Top-down conversion to the new culture.
2. Mutual goal setting and peer-level cooperation. Goals cut across all functions of the business. They are not split up and parceled out as merely separate, specialized performance measures assigned to separate functions.
3. The fostering of interpersonal skills and the participation of all employees in improvement—team meetings and team play.

And so was born the expression "Team Xerox." One of the most important elements of the creation of Team Xerox was the concept of the internal customer. That is, every Xerox employee should consider another employee as the recipient or customer of his or her work, then define that customer's requirement, and devise the corresponding supplier specifications. That process soon leads the participants to propose a measurement to see if the requirements of the customer are indeed being met.

An organization in which everyone is busy attempting to please internal customers is very different from one in which everyone is trying to please only the next higher level in the hierarchy. Nowhere was this cultural change pushed harder at Xerox than in the parts of the organization that dealt with suppliers.

It had been obvious in 1981 that Xerox had to both reduce material cost and develop much more competitive copier designs. Xerox needed the participation of suppliers to assist in refining copier designs as well as in refining the quality of the operations. Suppliers needed to become part of Team Xerox too.

Xerox formed multidisciplinary commodity teams to work with suppliers. Each team had a purchasing manager, but also engineers, materials analysts, systems specialists, and accountants. An especially important role on each team was that of quality engineer. These teams began working *with* suppliers to improve their operations.

Xerox also greatly reduced its number of suppliers—from 5,000 in 1980 to 300 by 1986. The weeding-out began by classifying each supplier not only on the basis of current cost or other performance, but by attitude, in three categories.

1. *Red* (Does not think improvement is necessary.)
2. *Yellow* (Slow to accept or manage change—marginal performer.)
3. *Green* (Willing to "go for it"—will try to meet world-class benchmarks—and strong enough to be likely long-term survivor.)

Some of the benchmark metrics Xerox thinks suggest operations proficiency include

- Ratios of functional costs (such as selling) to total revenue.
- Productivity (headcount per unit of output).
- Cost of order entry.
- Internal and external customer satisfaction rates.
- Internal and external defect rates.
- Service response time.
- Days on hand of inventory.
- Total manufacturing lead time.
- Time to develop a new product.

In other words, in the benchmarking of performance, financial profitability is only an indicator of health. More significant is an organization's ability to help strengthen companies it works with, and help make life miserable for companies it competes against. Table 1–1 summarizes a few Xerox performance indicators.

To narrow down the suppliers to those with whom Xerox wanted to have a long, close association, commodity teams visited the top management of prospective companies and asked for help comparing the subject company's performance with the best benchmark measures available from similar operations. Then, after an explanation of "excellent" manufacturing that concentrated on TQC, the suppliers were asked if

**TABLE 1–1**
**Xerox Performance Indicators**

| | 1980 | 1986 | 1987 Benchmark |
|---|---|---|---|
| *Quality* | | | |
| Supplier parts: defect rate Xerox assembly lines, (parts per million) | 10,000 | 450 | 125 |
| Internal: quality defects per 100 machines | 92 | 12 | 4 |
| *Cost* | | | |
| Direct material cost (index = 100) | 100 | 50 | 25 |
| Labor overhead rates | 380% | 180% | 150% |
| Material overhead rates | 9.0% | 3.1% | 2.5% |
| *Lead Times* | | | |
| Total manufacturing lead time (including longest lead time of any supplier) in months | 9 | 5 | 2 |
| Inventory, calender days on hand | 99 | 33 | 9 |

Notes: By 1986 Xerox had met or exceeded most of the quality benchmarks that they knew about in 1980, but by 1987 the best in the world had also progressed. Even the tremendous achievement of cutting materials cost by half still fell short of the world mark. It was, however, good enough to stay in the market.

they were willing to meet or beat world benchmark performance.

Xerox stopped spending excessive time with red and yellow suppliers, having found that a high fraction of Xerox staff time spent on suppliers had been consumed with red suppliers. Time concentrated on green suppliers builds toward Xerox's future; time spent on red ones is a waste.

With green suppliers Xerox rolled out extensive training in SPC, TQC, JIT (just-in-time), and EI. It worked. Defect rates in incoming materials dropped by 90 percent in three years.

Just as important, Xerox commodity teams learned the details of each supplier's production processes and management systems. They asked suppliers for suggestions on how to improve quality, reduce lead times, and cut costs. When this information was used in the design of new Xerox copiers, the cost of purchased material dropped by 50 percent as new designs came on line.

To accomplish this, Xerox assigned full-time buyers and production planners to its new product development teams. They reported to the chief engineer of those teams. Their key role was to work with the commodity teams and suppliers to draw out the involvement of suppliers as early in the concept phase of each new product as possible.

As a result of this effort Xerox copiers are now competitive again in the North American market. Had the changes not been made, Xerox would probably not still be in the copier business. In 1989, with the changes in place, Xerox won the coveted Malcolm Baldrige National Quality Award.

## TOKAI RUBBER INDUSTRIES, INC.

In recent years, Tokai Rubber has emerged as the leading supplier of industrial rubber products in Japan. The rise of the company has been remarkable. Almost as remarkable is the fact that the plant itself is spotless despite one of the most common ingredients in its rubber being carbon black.

Ten years ago Tokai Rubber was a sleepy, batch-oriented, production facility in a country where industry increasingly moved toward just-in-time production. Under a new chairman (from Sumitomo group) Tokai began a never-ending war on waste. Mr. Nozaki, then the new CEO, is almost the personification of Tokai. His is the guiding vision, but the workers are extremely involved through a goal-setting system known as policy deployment, sometimes described as "catch ball." Nozaki and other top executives make widely known their set of improvement goals for, say, the coming year. Then various worker groups consider the broad goals and determine what changes or projects they can undertake that will further the goals of Tokai Rubber. (This general approach is no longer a Japanese exclusive. It is being used by a few American companies too.)

Nozaki encourages the workers with numerous semimilitary metaphors. The main plant is the Komaki Plant, and Nozaki refers to the internal improvement campaigns as the "taking of the Komaki Castle," alluding to typical plots in Japanese Samurai movies (the Oriental equivalent of a grade-B Western). Nozaki possesses a remarkable ability to challenge and motivate the workers—an ability much prized among Japanese manufacturing executives strongly pursuing manufacturing "excellence" (not all are).

The "taking of the Komaki Castle" began with *total productive maintenance* (TPM), a term that is just catching on with American manufacturers. Essentially, TPM extends TQC to preventive maintenance. The operators assume major responsibility for knowing about their machines and caring for them. The objective is to improve the running of the machine, the time it is available, the process capability (statistical), the setup processes, and whatever else might make it into a first-class competitive piece of equipment.

TPM began with supervisors cleaning the machines, just to show the operators that it was useful and that they should do it themselves. (Such an approach does not touch the psyche of American workers the same way.) After a time, most of the workers were either embarrassed into participating in TPM or began to understand the reasons for it. Maintenance costs dropped, and machine availability time increased. The time required to change operator attitudes was about 18 months, but the delay was worth it because it began to reform the culture of Tokai Rubber.

Another prominent feature of Tokai's reform was Saturday training sessions for supervisors. The supervisors are paid overtime to attend training classes. Saturday training began about 1979 with a mandatory 10-week training course. Nozaki and other executives personally attended the courses in the beginning. Courses are extended and repeated now, and the executives still show up—a clear manifestation of the high regard that the top executives have for the nitty-gritty training of the people in the company.

Much of the training takes a show-then-do approach. Supervisors learn how to apply problem-solving methods to situations in their parts of the factory. Soon they gain enough experience to become teachers of the same techniques to the workers, which is the reason for the type and intensity of the training.

Another benefit of supervisory training is that people of the company meet regularly and form strong friendships. The problem-solving networks of the supervisors become very extensive. This approach is typical of the leading Japanese manufacturers, who believe that the heart of improvement in their companies is strengthening the "constitution" of the people in the company. After a time, supervisory training began to include various staff and white-collar personnel. As a result, networks are even wider, and the office workers have greatly improved their productivity also.

Of course, the company has maintained a strong quality program for a long time. In addition, they began JSK, their version of just-in-time production, in 1982. Factory production was reorganized into a number of cells. The initial *kanban*-style pull systems evolved into elaborate visible control boards inside the plant. Computers track some production data, but virtually all production control is done by simple visual means.

Figure 1–3 is a collection of several performance indicators kept by Tokai Rubber, but only a sample, not a complete

## FIGURE 1–3
## A Sampling of Performance Measures Used by Tokai Rubber Company
## from 1977 to 1987

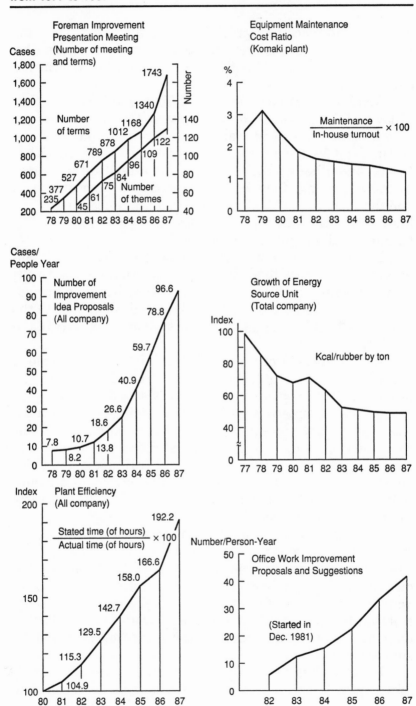

**FIGURE 1–3** (*concluded*)

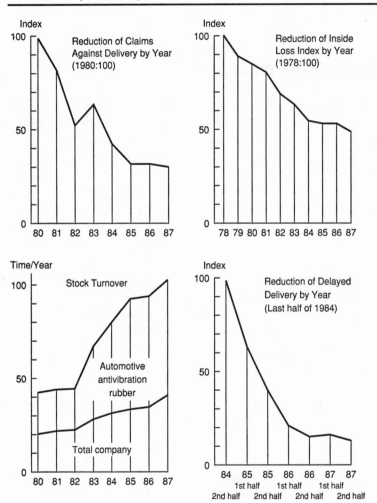

list. However, the exhibit suggests several performance measures Tokai considers important. Notice the extremely high number of improvement ideas per person per year. This measure is typical of Japanese manufacturers, and companies in other parts of the world do not come close to matching them on this measure.

Chairman Nozaki has a simple philosophy about manufacturing improvement. If the company tries to serve its customers well and is successful in removing waste from its operations, then profitability will take care of itself.

## LESSONS FROM THE CASES

It is easy to recognize strong similarities between the three cases, yet each is also different. Each company began from a different background. Each saw its threats differently. Each began its improvement program in a slightly different way. However, after reading a half-dozen or so such cases, the reader sees enough similarities in the plot to be bored.

The managements of each of these companies would not claim to be the best in the world. Some have the benchmark data to prove they are not the best. Part of their attitude is a continuous striving to be better, a recognition that no group of human beings ever reaches such a level of achievement that improvement is impossible. However, each of these companies is currently holding its own in tough, world-class competition. All three press continuous changes involving the three areas of Figure 1–1: people, quality, and process.

Each of the companies stressed at least four broad objectives:

1. Always working toward customer service and operations development targets.
2. Keeping quality supreme.
3. Reducing lead times by eliminating waste.
4. Reforming the organization; shifting responsibility to low levels of the organization and increasing the skills of all personnel.

Each of the companies also worked with its suppliers. Good performance inside companies was limited in value unless suppliers could perform almost equally well.

Readers of a marketing persuasion will probably think of this new paradigm as market-driven management. In a sense it is. However, concentrating too much on the market function, or any one function, makes a company unable to compete. The best marketing intentions are of little value unless the entire organization is committed to excellence of operations in order to support them.

Some of the measures that stimulate major reform are financial. When a company is losing so much money that the doors may shut, the situation presents a strong incentive for radical change. Yet, the financial measures in themselves suggest few changes in operations that are necessary in order to reclaim profitability. Nor is it necessary to be in dire straits to reform. Hewlett-Packard and Omark Industries (now Blount, Inc.), for example, began from visions rather than desperation.

Most of the measures that provide operational guidance in a company's reform efforts are nonfinancial. Some cannot be expressed on a printed page. Workplace organization inside a factory, for instance, is difficult to represent in the abstract. The detailed layout and other signals from the factory floor itself provide clues that show whether the plant is running properly; such visual clues often point out problems. Words and numbers cannot express as concisely the message conveyed by these visual patterns.

## CUSTOMER-RESPONSIVE MANUFACTURING: THE NEW PARADIGM

The term *paradigm shift* was originally used to refer to major changes in scientific thought, such as that produced in the world of physics by Einstein's theory of relativity. Recently, the term has become popular to denote major shifts in management or economic thinking—a philosophical revolution.

The management revolution itself can hardly be considered more than 100 years old. Before then few people regarded management as a human activity distinct from railroading, farming, soldiering, or whatever else they were engaged in. The new paradigm really represents a shift from the basic concepts that evolved from Fredrick Taylor's theory of scientific management. (The evolution did not go exactly as

Taylor envisioned it.) Most production people do not discuss paradigm shifts, however. They refer to a "new mindset" or a "change away from business as usual."

The reason for calling this change a paradigm shift is that thinking has changed in too many detailed ways to cover the entire picture with a few descriptions of new techniques—the alphabet soup of three-letter acronyms. Yet a brief discussion of the major changes can help put into perspective the overall shift taking place in measurement systems. The key differences between "then" and "now" are summarized in Table 1–2.

**TABLE 1–2**
**Contrasts in the Paradigms**

| "Business as Usual" (or Manager-Centered Manufacturing) | The New Manufacturing (or Customer-Centered Manufacturing) |
|---|---|
| A Company Is Assets<br>  –The company is a possession | A Company Is People<br>  –Assets are people; the company is people |
| Economy of Scale<br>  –Bigger is better | Economy of Time<br>  –Faster response is better |
| Managers Manage; Workers Work<br>  –Management and staff promote improvement | Workers Are Thinkers<br>  –Everybody as a team promotes improvement |
| Vertical Organization<br>  –Top down communication; walls between departments | Horizontal Organization<br>  –Multidirectional communication; next operation is a customer |
| Profit Is First<br>  –Cost trade-off thinking | Quality Is First<br>  –No-compromise, goal-driven thinking; customer satisfaction is number 1, and customer satisfaction = quality |
| Company-Centered Operations<br>  –Transaction-driven management | Manufacturing Networks<br>  –Improvement-driven, problem-solving teamwork |
| Performance Measurement for Control<br>  –Financial measures are dominant; more narrow, functional measures | Performance Measurement for Improvement<br>  –Customer satisfaction and noncost operating measures are more prominent; more measures of processes for mutual improvement |

## A Company Is People

According to the old way of thinking, a company is a collection of assets—capital equipment and tangible goods that can be assigned a value. The company is things, with exceptions. Sometimes the employees are described as a company's most important asset, but only in rare circumstances are they shown on a balance sheet under that heading. It is too hard to value the work of an employee, so accounting rules prevent such fuzzy entries in the books.

According to the new way of thinking a company is people. The Japanese manufacturing company, often taken as a model, regards itself as an aggregation of the people associated with it—mostly employees. The board of directors primarily represents the employees of the company.

In much of the rest of the world, the new direction is seen as "knowledge work" done by "knowledge workers." The definitions of these terms are somewhat vague, but they imply competence with computers. The new concept is not well-developed, but it is clear that the major resource of the world-class manufacturing company is the collective capabilities of the people who make it function. In a highly automated company, the major factor in success is not the equipment. That has to be versatile and change as the product and technology change. The major factor is the people who work on the process, the products, and the equipment—that employ the technology to serve the changing customer.

## Economy of Time

One of the surest ways to remove waste from operations is by reducing the lead time required for them. For example, suppose that one can reduce the lead time for material to go through an assembly plant. Previously, material made six moves and sat five times before leaving. Now it goes through in two moves: (1) incoming truck direct to assembly, and (2) assembly onto product. Thought of in this way, time reduction must obviously result in reduction of resources used. That is only one example.

The most difficult task is making this logic instinctive. Real breakthroughs in operational thinking are stimulated by thinking how to do more in less elapsed time. That forces simplification. It forces correct completion of work the first time. It forces no-trade-off thinking.

"Business as usual" is economy of scale, trade-off thinking, carried to a high art. The concept is that many costs are fixed or for practical purposes unchangeable. Therefore, to take the simplest case, the lowest cost per unit is derived by making as many units as possible with the same fixed cost. That is "run-as-a-routine" thinking. It is not imaginative improvement thinking. Unfortunately, economy-of-scale thinking is an unspoken assumption behind many kinds of performance measurement, including cost accounting systems.

**Workers Are Thinkers**

In the old paradigm factory workers mostly worked. Changes in equipment and systems were the business of staff and management, and a great distance in communication and status separated the two camps. Today the revolution in employee involvement is making steady advances. Perhaps this approach is not yet dominant, but more and more companies are at least listening to their workers. In the three cases presented, thinking workers are the backbone of reform.

The toughest role change is not that of the workers, although naturally many workers do not really want to be major contributors to improvement. The most difficult change is in the role of staff and management.

**Horizontal Organization**

Traditional manufacturing has been characterized by organization in vertical hierarchies, the general exceptions being small or new companies in which formalities have not had time to take hold. There are many problems with vertical hierarchies in the new manufacturing, beginning with the defect of excessively long, sequential decision making between departments. This arrangement is often summarily described as a *bureaucracy*.

The new paradigm is based on team management—a much more horizontal organization. Western companies everywhere are taking first steps toward this by negotiating major reductions in the total number of job classifications and by removing layers of middle management. This change is a huge subject with many ramifications. For instance, systems of performance measurement are obviously based in part on the way companies are organized. Change the organization and the systems of measurement also change.

## Quality Is First

Quality means giving customers what they need technically (even if they do not technically understand their needs) at an affordable price. Excellent quality means pleasing customers, giving them more than they thought they deserved. These platitudes translate into an enormous amount of work in attending to detail, identifying problems, developing solutions, standardizing results, and so forth.

The old paradigm is that you get only what you pay for; high quality costs more. The new paradigm is that if you pay attention to the customer in detail and do it right the first time, high quality costs less. It is less wasteful. (Of course, if quality means luxury and extra features, the old paradigm still holds, but reliable product service does not cost more.)

If quality costs more, trade-off thinking assumes that there must be a happy medium, so compromises are made. Defects are expected. One even hears about the "optimum" level of defects. But the new thinking allows no compromise. No other improvements compromise quality of result. No defect is considered acceptable.

## Manufacturing Networks

With business as usual every company is out for itself. Suppliers are not completely trusted, so communication with them is poor. Customer companies bid work more on price than on other qualifications of the supplier, and certainly they want to keep their options open—no sole-source suppliers are used if it can be helped.

As the Xerox case shows, much is to be gained from co-operation with a limited number of suppliers, and that is a current trend. Xerox has pressed further with it than most American companies. The reason for this close relationship with suppliers is to coordinate relatively complex development and production efforts. Although there are risks, the old method of working with suppliers—three bids and take the lowest—is too slow and too costly.

As a consequence the new paradigm has had a important impact on manufacturing networks consisting of multiple companies and operation locations. The former thinking was that if customers and suppliers behaved like competitors, the overall result would be more effective. But company after company is finding that this is not true. The new thinking is operations network thinking; the old thinking was business-centered, transactional thinking.

**Performance Measurement for Improvement**

Companies following the new paradigm all have an extensive improvement process in operation within the company. Measuring performance for improvement is different from keeping tabs on performance for routine reviews. The assumption of production performance under the old paradigm is often that the job of production people is merely to operate the system that staff and management have designed for them, so "keep-it-going" measures are used. Somewhat the same can be said for working-level persons in other functions of a company.

The decision to embark on a process of constant improvement means that most people should be measured on their contributions to improvement as much as or more than their performance of the routine parts of their work, which are often the parts they are trying to improve. More important, there is a difference in the attitude associated with being rated by a boss who primarily judges the recorded outcomes of work and mutually measuring the output of a process so that everyone as a team can improve performance. (However, problems of individual performance, and occasionally problems of discipline, do not go away. The new paradigm is not perfection.)

Another major difference is that this form of measurement places more emphasis on operating measures—customer satisfaction ratings, defect rates, on-time delivery—rather than on costs—dollar measures of resource use—although those obviously remain important. Some of the operating measures are close to the action, but all measures save the strictly visual ones are abstractions. Many are calculated, discussed, and used for decisions in locations far distant from the reality they are supposed to represent.

Good performance measures assist our thinking process, and the reality of the performance measurement process itself is within us, in how we develop the measures and in how we react to them. Performance measures are a record of prior human activity and a stimulus to new action. Otherwise they are of no more value than the coating of dust on a machine.

## NOTES

1. The case studies are based on articles and material which have previously appeared in *Target*, publication of the Association for Manufacturing Excellence:

*Harley-Davidson Motor Company:*
Thomas Gelb, "Harley-Davidson, a Company That's Taking a Different Route," *Target* 1, 3 (October 1985).
John Saathoff, "Maintaining Excellence through Change," *Target* 5, 2 (Spring 1989).

*Xerox Corporation:*
Larry Connorton, and Pierre Landry, "JIT Purchasing at Xerox," *Target* 3, 4 (Winter 1987).

*Eaton Automation Products Division—Watertown:*
Earl Hildebrandt, "Manufacturing Excellence through People," *Target* 4,1 (Spring 1988).

*Tokai Rubber Company:*
Per Johansen, "Taking the Komaki Castle," *Target* 4, 4 (Winter 1988).

# CHAPTER 2

## OUR MEASUREMENTS MODEL OUR MINDSETS

Performance measures are derived from our true goals. Performance measures help define our goals. However, performance measures do not in themselves attain goals.

For example, most manufacturing plants have water meters, but the water meter does not curb the waste of water. People can curb it, and then only if they believe that restricting water use is a beneficial goal. Many measurements have little effect on performance, but those tied to the vital few performance goals have a profound effect.

Much of the discussion of performance measures by companies changing to the "new paradigm" is a surrogate discussion of goals. Discussing changes of goals and values directly may be too emotional. However, just stating a goal as a performance measure may be insufficient communication.

The late E. F. Schumacher, author of *Small Is Beautiful*, illustrated this point by recalling that he, as a young statistician, worked on a farm during the war. Each morning he was instructed to go to the top of yonder hill and count the cattle in the field. Every morning he returned with an accurate count—even on the morning he discovered one cow dead of a disease. Only then did he realize that his real responsibility had been not just to give a count each morning, but on the prior morning to have added, "But one cow is a bit mimsy, sir."[1]

## GOALS OF MANUFACTURING
## IMPROVEMENT

The goals of the "new manufacturing" are intended to progressively develop an extremely strong competitive organization through the development of people, collectively and individually. These goals are quite different from the sales, cost, and profit goals that dominate business-as-usual thinking. The difference in goals creates many problems with performance measurement.

The goals emphasized during an improvement process depend on the history of an organization and what it must do to compete; that is, to continue to satisfy the reasonable demands of its customers in the future. Correct goals are critical to survival in highly competitive manufacturing.

At the beginning of the 1980s both Harley-Davidson and Xerox offered their customers high-end prices with low-end quality. Therefore, both companies had to greatly improve quality internally as well as from suppliers, and both companies had to drastically reduce the waste of resources. Harley-Davidson was more vertically integrated, so its attack on quality and waste at first focused inside the company in order to change its culture; but Xerox fabricated only a few components of its copiers, so the company's improvement strategy had to bring results from suppliers relatively quickly.

By contrast with Xerox and Harley-Davidson, Tokai Rubber in 1980 was not in a survival crisis. It had already been working on quality improvement for some time. Its more serious competitive shortcomings were poorly adhering to schedules and wasting resources. To improve dependability and cut waste, Chairman Nozaki promoted Total Productive Maintenance to accomplish these goals through better operating readiness and process capabilities of the equipment. In a sense, the logic of TPM was almost an extension of that which had been used earlier to promote quality, but TPM still required extensive training.

Among companies engaged in a continuous improvement process, the process boils down to a never-ending program of developing people. Unless strategic considerations dictate otherwise, most companies prefer to start with quality improvement since the disciplined problem-solving associated with quality is fundamental to almost all improvement. The objectives for people-centered improvement, sorted into five categories, are:

1. *Quality:* defined as customer satisfaction with products and services at a satisfactory price. The definition of quality as conformance to specification is not sufficient, because part of customer satisfaction is determined by the specifications necessary to create sale-to-salvage customer satisfaction. The definition of customer also extends to "secondary" customers: these may be dealers, service personnel, or simply people exposed to a product in addition to the primary buyer and user. In the sense of environmental safety, everyone on Earth is a "tertiary" customer.

2. *Dependability:* consistently performing at the time scheduled or promised. On-time delivery or immediate availability to the customer are the result of dependability; but to attain good delivery performance, production should finish when needed, engineering changes should execute as planned, service representatives should arrive when needed, or suppliers should deliver on time.

3. *Waste Reduction:* defined as elimination of any unnecessary material, time, space, equipment, or activity, or as elimination of superfluous activity. This definition is well-known as the major objective of JIT. Inventory reduction is a small part of the overall intent.

   Said differently, waste reduction is no trade-off thinking in reducing the use of resources, both tangible and intangible, whether easily enumerated or seemingly nebulous. Reducing waste is typically a simplification of the process. Simpler processes are easier to operate dependably and consistently, and consistency in lead times or other measures is an indicator of quality performance. Goals and measures interrelate and build on each other.

4. *Flexibility:* defined as reduction of the lead time to make any kind of significant change, or as responsiveness to change. Thoughtful reduction of the lead times to perform repetitive activities is part of eliminating waste. Reduction of lead times to increase production, to change production mix, or to introduce new products is a step beyond that.

   For example, converting to cell production with quick changeover capability addresses the problems of material transport delay. However, to be truly maximum in flexibility, the layout and the available tooling in such a cell should be rapidly convertible to a radically different production mix. Beyond that, the company should be skilled in quickly devising a new cell arrangement to produce a new product. That skill will help introduce new products sooner.

   An example of flexibility is shown at a Japanese auto assembly plant that can, over a weekend, change line speed to any rate within ±15 percent of the mid-range. With this capability, developed through team coverage of line zones and team responsibility for line-side revisions, plus variable work days and work weeks, the assembly plants can track the changes in market volume and mix closely. The skills that allow the plants to do this are also valuable for quickly transforming the plants at model-change times. (And the plants are still not as flexible as they want to be.)

5. *Innovation:* defined as successfully introducing new technology or service to a market—origination of useful new practices. Companies can be flexible while merely being fast-followers, and a company that is innovative but inflexible, may find its originality smothered by fast-followers. Being an innovator is desirable. A reputation for successful innovation is useful, and innovators enjoy a brief period of high margin without competition.

The more successful companies have recognized that their improvement processes center on people. The preceding objectives are for improving people, and through them, for improving organizational performance in many ways.

In the past, companies could afford delays when bringing new products to market if they enjoyed big economies of scale over rivals. A General Motors or a General Electric could overwhelm upstarts by spreading their large, fixed costs over huge market volumes generated by marketing power. As mature markets fractionate into niches and medium-size companies remove the waste from their operations, this advantage is fading. Product and process development is becoming a part of an overall "rat race" in which the winning companies are those who can quickly master the problems of delivering quality to the customer. The ability to do that is built through teamwork for operating excellence—market research, engineering, production, field service—all functions working as teams.

Companies cannot expect to succeed in this new competition just by inferring other objectives through cost measures. They must strive to eventually have innovation *with* quality, dependability, waste reduction, and flexibility—performance excellence.

Every company progressing through these objectives by continuous improvement has an *improvement process*. Xerox has a nine-step improvement process. Ford has an eight-step process dubbed "8-D." Most of the company-specific improvement methods are derived from the granddaddy of the genre, the Deming Circle (or Deming Cycle): Plan, Do, Check, Act; or PDCA. The Deming Circle is an adaptation of the scientific method, which most children learn in some form in elementary school, and that later recedes into dim memory.

An improvement process is a collective, cooperative effort to apply scientific logic to business and manufacturing situations. The logic is useful for large problems or small—boardroom to locker room. See Box 2–1.

A major thrust of improvement by scientific means is with management by fact and conclusions from data. Most managers in manufacturing consider themselves to be highly rational and frequently base their thinking on some kind of data. Unfortunately, pseudoscience in the area of improvements come easy. One of the more common ways to short circuit logic is by weighing all improvement moves by an official cost system without factoring in the biases of the system. Data should pertain to the actual phenomena of interest and to the strategic objectives. This aspect of measurement creates a great deal of frustration.

Good *measurements* for improvement processes:

* Assist making general goals specific.
* Compare actual status to goals.
* Suggest where to direct effort.

---

**BOX 2–1**
**Improvement Processes**

To almost everyone never before exposed to it, the idea of a stage-by-stage, goal-driven process of improvement is absorbed slowly, and sometimes fitfully. It is an emotional acceptance as much as an intellectual one. In fact, the intellectually quick may dismiss the whole approach as nonsense—easily understood and elementary. The words with which to express improvement processes cannot impart something never before experienced. An improvement process is a *people improvement* process, and among the people to be improved are the leaders of the process.

When a company—and an industry—has an improvement process, working life consists of improving yourself and making improvement in every process you deal with. Often called "managing the process," making changes by this approach is *the way of running the business*. The concept of an improvement process is old—so old that its origin is lost in history. It has been practiced throughout the world.

Unfortunately, around 30 years ago Japan was famous for poor quality. In the absence of an improvement process based on scientific method, and in which every person in the company would participate in order to cover every detail, traditional Japanese management never rose above mediocrity. However, in the past 30 years Japanese companies refined the use of a scientific method in practice, by all employees to advance operational practice to a world-competitive level. The most common codification of the scientific method used in Japan is the Deming Circle: Plan-Do-Check-Act.

Plan: Search for problems. (Are we working on the right problems? etc.)

Do: Analyze. Ask why five times. Find root causes. Propose a solution.

Check: Try the proposed solution. Collect data. Verify that it worked.

Act: Standardize the solution in practice until better can be found.

Simple as the Deming Circle may seem, few people have the discipline to practice all four steps completely unless they make a strong effort to learn how. It is the development of patience, persistence, and attitude that is more difficult than understanding so simple a method.

In the United States many companies have also adopted the same kind of thinking, sometimes using the Deming Circle, sometimes developing their own variant of the logic. An example shown below is an amalgamation of a "customer satisfaction process" used by several companies.

1. Identify customer.

Who is our specific customer, or what is our market niche? What are the true needs of the customer, whether the customer realizes them or not?

2. Define the customers' specifications.

In detail, establish the nature of the product, service, and overall condition that is necessary to please each customer. Determine how to measure customer satisfaction.

3. Define the existing process for serving the customer.

Understand the process: Product flows, information flows, quality checks, capabilities of equipment and procedures, capacities, actual understanding by people, and so forth. (Many companies do not understand their processes, especially when they extend across many organizational boundaries.)

4. Propose how to change the process to meet the customers' specifications.

Gather ideas. Formulate them into a plan for a trial change or pilot program. Devise a plan for executing the change. (Most companies have trouble developing skill in enough people so that they could actually make the detailed changes. *Ideas* are comparatively easy.)

5. Follow up to ensure customer satisfaction.

Try the proposed change. Collect data to see if the gap between customer expectation and actual performance has narrowed. (This is easily skipped in the rush to move on to other things.)

6. Ensure continued customer satisfaction.

Standardize the solution. If possible, devise fail-safe ways that prevent old problems from recurring. Ensure that new knowledge is actually understood and practiced by the people doing the work. (Poor standardization is the cause of many chronic customer complaints.)

This logical framework appears so simple as to be boring. However, that is its beauty. Anyone can fundamentally understand it. The framework is also very broad. There are many different kinds of customers. Some are for products and services. Others are third parties, for example the general public, which needs to be served by preserving a clean environment.

Simple as it is, this kind of logical framework is the basis for organizational progress in manufacturing, provided that the goals of progress can be clearly defined. This framework is not that of financial planning or control; nor legal thinking; nor make-and-sell thinking. *It does require* the participation of all employees, so the style of management must encourage that.

> For example, if one concentrates only on profitability as the goal of the company, the actions of one person or a single financial transaction can lead to a higher profit than the work of thousands of other employees (or so it can appear), but if the goal is operational excellence serving many different kinds of customers of a company, *every* employee must be engaged in that.

## ESCAPING THE COST JUSTIFICATION TRAP

With business-as-usual thinking, manufacturing improvement means cost reduction, because profit is the overriding goal. For example, a drastic 10 percent across-the-board cut, done mindlessly, results in cutting tooling budgets at the very time when tooling needs rework to prevent scrap. Managements at their worst do not have time to think through the operating cause-and-effect linkages.

More prosaic versions of cost-cutting improvement are generally consigned to engineering departments. If high enough payback can be shown, perform a number of small projects: overhaul a machine tool or rebalance an assembly line; simplify the design of an old product to reduce cost through value engineering. And a classic one, buy in higher volumes to obtain a lower price.

A company finds itself in the cost justification trap if each cost reduction project's returns are measured independently of most other projects. Only those projects that pass a high financial hurdle are adopted, on the assumption that effort is better spent elsewhere. Cost savings and return on investment dominate judgment of improvement, often superseding even strategic considerations, and the long-term development of people is hardly considered. Improvement strategy is caught in a cost justification trap.

Staff engineers prefer big-step projects. These are often done in conjunction with new marketing thrusts, so strategic considerations are more prevalent. There is more money, more visibility, and more technical excitement. As an example, at the beginning of the 80s, Xerox committed itself to the future by building new assembly lines designed to improve line balance at Webster, New York, complete with an automatic storage and retrieval system to feed it material. Technically, it was state of the art. Financially it could be justified by providing a lower unit cost and a good return on investment based on expected future volumes. That was before Xerox gained a vision of the new manufacturing strategy.

A classic example of cost justification thinking is the practice of automotive companies to revamp old assembly plants with state-of-the-art equipment when a new model replaces an old one. The anticipated volume of the new model provides the ROI justification; it would be hard to justify improvement changes for the existing model.

Financial justification is embedded in the economy of scale, and the "substitute-capital-for-labor" mindset of mass production. Cost justification truncates thinking, whether the calculations consist of only a simple payback method or involve complex, discounted, cash-flow models. The problem with the thinking is less with the financial models than with the narrow view of the business.

For example, an analysis of the 118-year history of the Boott Cotton Mills of Lowell, Massachusetts, followed the mill through many economic cycles, sometimes losing money, sometimes making a fat return. Addicted to financial justification, owner after owner regarded the mill as a money-in, money-out proposition. Major investments in the mill could only be justified by the prospect of big profits. Improvements in operations could not be considered, so a dying business could never rescue itself.[2]

This story illustrates the problem of a certain mental perspective rather than some defect in cost analysis. So long as the major purpose of an operating-level system is to extend no further than making money for the ownership, the cost justification mind-set follows: every incremental change should be incrementally justified. The mind-set is not necessary and is even counterproductive to its own ends. For example, according to Dr. Kano, the return on total assets of Japanese companies winning the Deming Prize during the 1970s (a sign of a significant quality improvement process) was about double that of the average Japanese manufacturer.[3]

Guiding decisions exclusively by cost trade-off has never been good management practice. Accomplished managers have always thought more deeply. But cost-dominated improvement decisions have often prevailed.

Reliance on financial justification was not a disadvantage so long as well-engineered capital substituting for labor could result in competitiveness. Strategic operating capabilities and the development of *all* people to undertake them are becoming the new determinants of success. That is, poor use of advanced equipment in a process without continuous improvement is a liability, not an asset. Integrated development of people and processes make the difference.

At the same time, quantification of resource use is necessary for many reasons. Measures of resource use may be monetary (costs) or ratios, such as productivity measures, simple eyeball estimates from process flowcharts, or direct observation of processes.

By the new philosophy, all employees at all levels in a company should always attempt to improve the processes of the company—an ideal never completely realized. Justification of an improvement activity depends on whether it furthers an important goal at a reasonable cost. The view of the enterprise and the approach to decision making differ from business as usual.

The financial view of business as usual is that manufacturing can be reduced into an ongoing series of decisions whether the probable gain from each improvement activity, independently considered, will exceed the expenditure for it. Exceptions are made, but before proceeding with most proposals, the burden of proof is whether it will show an acceptable gain, generally by comparing net investment versus operating gain by simple means or complex.

The decision making of the new manufacturing is goal-directed. The method of advance is primarily through improving people capabilities. Improvement activities should form an overall pattern of performance improvement, so the merits of each one do not necessarily have to be narrowly evaluated, independently of the rest. Each will be undertaken if resources are available and costs are not prohibitive. Thus, this logic differs from business-as-usual where the burden of proof lies.

## OVERALL IMPROVEMENT BUDGETS

However, no company can undertake every change at once. Resources are limited; choices must be made. One approach is to set aside an overall budget to be used for improvement. Undoubtedly, some of the budget will be misspent because people have to learn through mistakes, but in total, the budget should advance the company toward its improvement goals.

Staying within an overall improvement budget can be done with minimal attention to a detailed budget. A production facility that is trying to reduce lead times of material should attempt to reduce setup times, for instance. The number of setups that can be executed must stay within the overall work time of operators, setup specialists, and others. Start by posting, for example, how many process setups take place per month before improvement. Set a target to increase the number *without adding any people*. Track the progress. If, in addition, all machine and tool modifications are done in-house, big capital spending is restricted. Advancement is through waste reduction and the increased skill of the employees themselves. Although everyone should stay within overall budget limits, the spending is limited by the improvement policy itself.

In the new manufacturing, the way to make money is to stop worrying whether every *detailed* action contributes to profit. Worry whether the company has a competitive strategy and has the necessary improvement process in place so that the people of the company can execute the strategy well: these serve the targeted customers better than most efforts. Improvement goals are primary; financial goals follow them. Difficulty changing that mind-set in practice is one of the causes of problems measuring performance.

## ESTABLISHING IMPROVEMENT GOALS

Goals begin with development of long-range competitive strategy, and a vision of how to attain the strategy. Competitive strategy is a subject unto itself. From it, a vision should set the overall direction of improvement processes, but visions are not necessarily measurable goals. It should be possible to measure progress against goals.

A company's vision and philosophy are often articulated in a vision statement, or mission statement. Those statements based on the new manufacturing feature customer service as the centerpiece, but in addition the statements are crafted from attention to customer needs, familiarity with technical possibilities, and knowledge of operational performance excellence.

While keying a company's vision to customer satisfaction rather than profitability is more socially defensible, the supremacy of customer satisfaction can also be challenged. Critics can point to worst cases of customer satisfaction as selling booze to juveniles, guns to criminals, and wildlife refuges to strip miners. One has to assume that customers are socially responsible (not all are), and that attentive companies can technically define customers' needs better than the customers themselves.

To do anything practical, a company must emerge from this philosophical quagmire with a cohesive strategy of improvement and goals that can be measured. A company's operations should become better in fulfilling its mission when measured by indicators of six kinds of performance: human development, quality, dependability, waste reduction, flexibility, and innovation (H/Q/D/W/F/I). Many of the items measured are merely enablers of performance, and not performance itself. In any case, measurements are no substitute for real performance. Reality in performance, regardless of measures, is whether a company can sustain preferential status with its customers.

How does measurement fit within an overall improvement process? First, the goals of the process must be guided by the needs of customers and the areas in which the people of the company must improve—a competitive improvement strategy. Areas for improvement need to be described *with* measurements (customer ratings, customer returns, setups per month, etc.) To focus attention, objectives may be set as targets positioned according to the measurements.

Targets, and the measurements that evaluate performance toward them, need to relate to something tangible: customers, products, processes, and people. Working to achieve something tangible is more pleasing than striving for an empty number. An arbitrary cipher is what an ROI-rate hurdle means to most employees.

Working their way into the new manufacturing, organizations change their performance measures to reflect their new goals. For example, the Capacitor and Power Protection Division of General Electric revised its internal performance measures based on a survey of customers. Capacitors and surge protectors are high-volume, low-margin products. Management had earlier concentrated on measuring volume of output and its yield of acceptable products. The goal was to maximize both numbers, but that stimulated everyone to run processes at top speed and maximize output by accepting marginal-quality production.

The customer survey indicated that customers were interested in the lowest *price* to them, not GE's apparent internal cost; and they were interested in *on-time receipt when they wanted*, regardless of whether the plant had shipped on time by an original promise. The result was not surprising, but it caused GE to rethink its goals and measures for production and shipping. Production measures were revised to emphasize *process quality*, upgrading the process to "Do it right the first time." Other measures concentrated on making and shipping *what* customers wanted *when* they wanted—to increase responsiveness and decrease lead times.

In this way, the Capacitor and Power Protection Division continued to reassess production goals and performance measures to align them with the operating requirements for the needs that the customers considered critical. Results so far are positive from customers, and the division still has many improvements it can make.

Effective improvement goals and measurements must also motivate. Motivating goals are accepted as challenges. For example, in the early days of quality improvement, Motorola's top management decided to peg their quality goals to a *tenfold* reduction in defect rates rather than a more modest, "realistic," 10 percent or so reduction per year. The goal became meaningful when the company began to coach its employees on *how* to reach it. Without realizing it was possible, the employees would have ignored it. The tenfold goal itself reinforced the fact that business as usual was no longer acceptable. A major change was necessary.

The early tenfold-reduction goal prepared Motorola employees for a second major challenge set later, the six-sigma quality goal. The goal roughly is that each Motorola process should have so little variance compared with the design specifications that there would be a defect only once in 3.6 million units.

## THE INTERRELATIONSHIPS OF OPERATING GOALS (OR WHY REDUCE SETUP TIMES?)

Improvement goals need to be set in the right place in an organization. This will be discussed further in Chapter 5, but the setting of goals begins with an understanding of the interrelationships between various goals. The interrelationships are based on the specifics of each company's products and processes; but there are many similarities, so a general discussion is possible.

As an example, try to think of all the reasons why setup times should be reduced. The reason for this example is that by business-as-usual reasoning, reducing setup times is seldom considered as an improvement goal.

The first reason most people think of is that reducing setup times allows reducing lot sizes, in turn reducing the materials throughput times for processing. (Those who have never heard of JIT will say that it enables the increasing of run time and increasing of efficiency.) However, many more reasons exist for reducing setup times.

Start with quality reasons. Setup times cannot be reduced much if tooling or equipment is in bad repair, so that special time for compensating adjustments is necessary with each setup. Perhaps the problem is material. For example, great variance in the thickness of sheet metal will increase the difficulty of setting up a press to obtain a correct part. In fact, if reducing setup times forces an examination of why a setup procedure cannot be quickly standardized, a process capability study could result. That can trigger a full investigation of quality, beginning with why the standards for the parts being produced are set as they are.

Presuming that a setup is refined until it is capable of producing a good part on the first machine cycle, another question is how long a run could last until the process should be checked. That is, what is the fail-safe quality time that can ensue after a setup?

Then how should a setup be checked and in what way? This question raises the issue of measuring gages and instruments and who is trained to use them. The problem of instrument calibration and repair also arises. One reason for a long setup is that either the people at the machine do not know how to use an instrument to check it, or if they do, the instruments at hand are broken or read erroneously.

Another reason setups may take a long time is for tooling to be checked and refurbished, and doing so takes a long lead time. This possibility brings up all kinds of issues of cross-training, priorities of work in tool shops, communication of tooling needs, and on and on and on. Interactive possibilities seem endless, and this same kind of what-if chain of reasoning can start all over again by analyzing another kind of improvement activity, reducing the nonavailable-machine time in a plant, for instance.

Both reducing setup times and increasing machine availability through better maintenance are represented more than once in the cause-and-effect diagrams in Figure 2–1. By contrast, a diagram of a simple version of the DuPont formula, familiar to those who have studied finance, appears also in Figure 2–1. The DuPont formula represents one diagrammatic way to illustrate business as usual, ROI thinking about improvement; the cause-and-effect diagrams represent some operating interrelationships considered from the viewpoint of the new manufacturing.

The DuPont diagram is symbolic of a financial system of thought. The primary interrelationships of an operating company are conceived as the calculus of a return on investment, starting from the detail level and working up to the grand measure itself.

The cause-and-effect diagrams in Figure 2–1 symbolize operations thinking: how people and physical phenomena fit together to create an effective result. The cause-and-effect diagram is one of the old tools of quality analysis that helps to sift through myriad possibilities to find causes not previously considered.

Both forms of thinking have their merits. Most persons experienced in manufacturing have met manufacturing owners who knew the customers and operations thoroughly, but had no sense where the cash went or how to price to make a profit. On the other hand, those whose mental processes seldom leave the cost-profit channel may have difficulty seeing the ideas behind the new manufacturing—which are really not very new in origin—without thinking rather reflectively about operating linkages.

A mental fixation on the goals of the new manufacturing plus concentration on operating interrelations are useful for creating improvement goals. It is tempting for those at the top of an organization to set very detailed improvement goals, issuing long lists of things to accomplish. Better to set a few well chosen goals at the core of the organization, then let those close to the work determine the specific detailed goals of their operations that will contribute to overall accomplishment.

In 1981 Tokai Rubber's Chairman, Nozaki, set an improvement goal of reducing unplanned downtime while at least maintaining gains previously made in quality (defect rates). Exactly the improvement activities that might contribute to that goal under the banner of Total Productive

## FIGURE 2–1
### Different Performance Measure Diagrams

1. DuPont Return on Investment Formula (Financial Thinking)

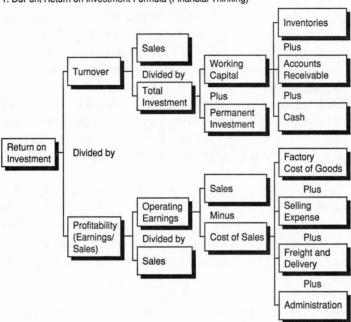

2. Customer Satisfaction Cause-and-Effect Diagram (Operational Thinking)

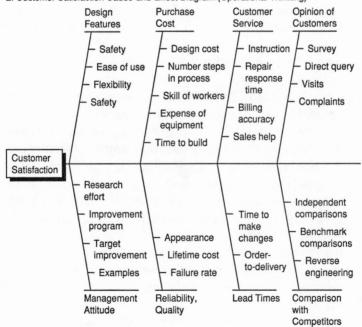

**FIGURE 2–1** (*concluded*)

3. Internal Quality Cause-and-Effect Diagram (Operational Thinking)

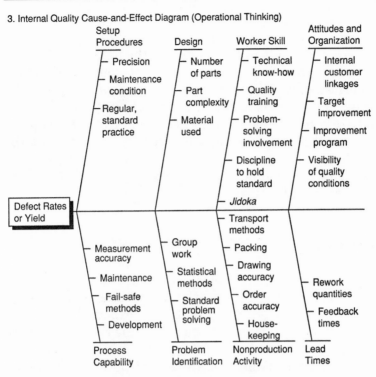

| Setup Procedures | Design | Worker Skill | Attitudes and Organization |
|---|---|---|---|
| – Precision | – Number of parts | – Technical know-how | – Internal customer linkages |
| – Maintenance condition | – Part complexity | – Quality training | – Target improvement |
| – Regular, standard practice | – Material used | – Problem-solving involvement | – Improvement program |
| | | – Discipline to hold standard | – Visibility of quality conditions |
| | | – *Jidoka* | |

Defect Rates or Yield

| | | – Transport methods | |
| – Measurement accuracy | – Group work | – Packing | |
| – Maintenance | – Statistical methods | – Drawing accuracy | – Rework quantities |
| – Fail-safe methods | – Standard problem solving | – Order accuracy | – Feedback times |
| – Development | | – House-keeping | |

| Process Capability | Problem Identification | Nonproduction Activity | Lead Times |
|---|---|---|---|

4. Production Lead Time Cause-and-Effect Diagram (Operational Thinking)

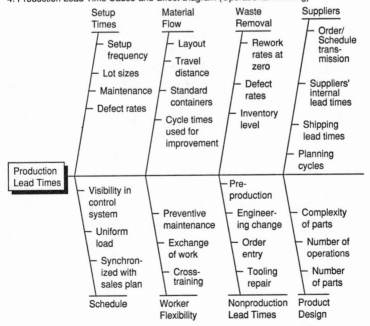

| Setup Times | Material Flow | Waste Removal | Suppliers |
|---|---|---|---|
| – Setup frequency | – Layout | – Rework rates at zero | – Order/Schedule transmission |
| – Lot sizes | – Travel distance | – Defect rates | – Suppliers' internal lead times |
| – Maintenance | – Standard containers | – Inventory level | – Shipping lead times |
| – Defect rates | – Cycle times used for improvement | | – Planning cycles |

Production Lead Times

| | | – Pre-production | |
| – Visibility in control system | | – Engineer-ing change | – Complexity of parts |
| – Uniform load | – Preventive maintenance | – Order entry | – Number of operations |
| – Synchron-ized with sales plan | – Exchange of work | – Tooling repair | – Number of parts |
| | – Cross-training | | |

| Schedule | Worker Flexibility | Nonproduction Lead Times | Product Design |
|---|---|---|---|

Maintenance was left to those close to the action as they began to learn what to do. Those Japanese companies, and a few American ones, who engage in *hoshin kanri*, or policy deployment, annually issue five or six major overall goals or directives from top management at the center of the company. ("At the center" is becoming more descriptive than "at the top.") Others respond with proposals for specific projects and changes, often including tests with those proposals by which their success will be measured.

## HOW DO WE KNOW WE ARE IMPROVING?

Certainly one measure of overall success is financial. Has the company's profit grown, given the delay time before the effects of an improvement program can be expected to pay off? That varies from company to company. Some can reverse cash deficits quicker than others.

However, there is no guarantee that performance excellence will result in a windfall to the ownership. The first objective is merely to survive. If a company is matched against very tough competitors, its operations can be very good and still not provide great return for owners. The customers receive nearly all the benefit. An industry full of tough competitors cuts each others' margins while providing, in total, higher quality service to the customers.

Another measure of success is market share. A successful company at least keeps the market share it wants of the market segment it serves. If it intends to grow, which is not always part of the strategy, market share may increase. For many companies, a good sign of competitive strength is taking customers from competitors who retain strong operating capabilities. (A company occasionally gains by default. Their competitors slip so badly that they receive business as a "gift.")

There are three methods of comparison to determine if operating capability is improving: (1) measuring against the same organization's prior history, (2) benchmarking performance against competitors' or against others with similar operations, and (3) comparisons with ultimate or fixed standards of performance. A book on each of the three methods of comparison is possible by elaborating on the details.

## Comparison with Prior Performance

The data for this type of comparison is easiest to acquire, for it all comes from within the organization. Even so, a typical problem of comparison is that measures that seem important after a continuous improvement process starts had never seemed important enough to keep before. For example, a record of how many setups a department performs each month does not seem very useful unless increasing the number becomes a performance goal.

Historical self comparisons are most common. Table 2–1 shows some typical self-progression data from several companies.

The obvious weakness of comparing against one's own prior performance is that the amount of improvement gives little clue whether the current level of performance is good. If a machine dedicated to making only one part is converted to make two parts, the percentage of improvement in number of setups per month would be infinite, but such a comparison is little indication whether the improvement is effective or competitive.

Much the same argument can be made about all comparisons with prior performance of the same organization or process. To judge whether improvement is effective, more than self-comparison is necessary. Some self-improvement comparisons are as meaningful as a four-pack-a-day smoker cutting to two. One cannot say there was no progress, but the progress may have stopped far short of being truly effective.

## Comparison with Others—Benchmarking

The basic idea of comparing performance with others is very old. Financial ratio comparisons between companies of similar type is so established that companies such as Dun & Bradstreet and Standard & Poor's have long made a business of compiling and selling the data. What is new in the last decade

**TABLE 2–1**
**Comparisons of Data Gathered for Performance Evaluations**

| Eaton Automation Products Division | | | Davidson Instrument Panel | | |
|---|---|---|---|---|---|
| | 1984 | 1987 | | 1983 | 1987 |
| Supplier defects | 1.2% | 110 ppm | Scrap | 5% | 2% |
| PC board touch up | 65% | 20% | Returns | 3.7% | .8% |
| Total inventory index | 100 | 35 | Number of suppliers | 100 | 35 |
| On-time shipping | 50% | 85% | Lost time accidents | 44 | 0 |
| | | | Total inventory | Weeks | 7 Days* |
| | | | Ace teams | 0% | 16%† |

| Toledo Scale | | | | |
|---|---|---|---|---|
| | 1985 | 1986 | 1987 | 1988 |
| Historical production quantities | 240,000 | 310,000 | 342,000 | 435,000 |
| Historical first pass yield | 78% | 85% | 89% | 92% |
| Historical nonconformance returns | 9.8% | 8.6% | 3.6% | 2.2% |
| Historical scrap and rework | 1.8% | 3.1% | 1.0% | 0.7% |
| Historical guarantee (warranty) expense | 4.1% | 3.1% | 1.8% | 1.3% |

| Globe Metallurgical Inc. | Index of Operations at 1986 Standards | Actual 1988 |
|---|---|---|
| *Beverly, Ohio, Plant* | | |
| Direct Labor | 100% | 30% |
| Salaried Employees | 100% | 30% |
| *Selma, Alabama, Plant* | | |
| Direct Labor | 100% | 63% |
| Salaried Employees | 100% | 56% |
| Total Employees | 100% | 37% |
| Total Sales | 100% | 128% |
| Productivity: Sales/Employee | 100% | 334% |

\* Warehouse eliminated
† Goal is 100%

All data has previously been published in *Target*, publication of the Association for Manufacturing Excellence.

is use of a broader range of comparative measures with other manufacturers. The best known company for benchmarking in the 1980s was Xerox.

Comprehensive benchmarking takes place on at least three dimensions: (1) financial and cost data, (2) product data, and (3) operating performance data. Any type of comparison can become complex, so it is easy to believe that a great deal is known when actually a comparison has been made using a few summary points.

Financial ratio comparisons of publicly-held companies, so far as they go, are done with data that is relatively easy to obtain. Much can also be learned about a competitor's products just by buying some and performing reverse engineering—test, disassembly, and analysis. However, most operating data can be obtained only if the company being benchmarked wants to give it up. The prevailing way to obtain data on competitors seems to be through informal exchanges, although excluding sensitive comparisons such as profit margins, proprietary process performance data, and product development data.

Once engaged in an exchange network, to obtain data about comparable operations elsewhere is less of a challenge than to use it constructively. Data does not do anything by itself. Companies like Xerox, that are experienced in benchmarking, use the data to keep abreast of competitor's capabilities, and especially to peg their improvement goals against operations they have discovered to be "best in class." The American benchmarking companies can be spotted because the phrase best-in-class (BIC) will be in common use. Improvement processes aspire to push organizations into BIC on various measures of performance.

Keeping masses of data on various operations around the world is unnecessary. With experience, benchmarking companies learn to track the more meaningful pieces of data from comparable operations that are at or near BIC. A brief overview of the Xerox system of benchmarking is in Box 2–2.

Probably the greatest value of benchmarking comes from setting best-in-class improvement goals after studying the product design and operational capabilities of competitors

**BOX 2–2**
**Benchmarking According to Xerox**

Xerox defines competitive benchmarking as "the continuous process of measuring our products, services, and practices against our toughest competitors or those companies renowned as the leaders." Observe what companies are doing now and project their performance in the future. The current benchmark is the very best performance of any given activity by any company anywhere in the world. All current benchmarks that exceed the performance of Xerox or of its suppliers become goals for future achievement.

Just as important as measuring the performance of other companies is determining how they achieve it. Whenever it makes sense, Xerox adapts the best practices of others to its own use. Benchmarks are periodically updated to provide new goals and new insights. Competitive benchmarking is not just taking periodic measurements. It is embedded in the ongoing Xerox management process. Xerox considers the competitive benchmarking process to have five phases:

**1.** Planning. *What to benchmark?* In the case of direct competitors, almost anything can be benchmarked, although with experience, companies use only the more vital measures. The same holds for suppliers and their prime competitors. In the case of renowned companies, competitors can benchmark operations of interest such as order-entry-to-customer-delivery activities when the outstanding feature is distribution, for instance.

Items to benchmark include throughput times, lead times, key defect rates—descriptors of quality, flexibility, cost, and delivery plus customer satisfaction. Xerox has a database on many kinds of companies covering the classic cost elements: wages, benefits, efficiency, energy cost, cost of capital, depreciation, taxes, and all the categories of overhead, including profit. Overhead allocations are sometimes known down to the workstation.

The question is, which data is obtainable, reliable, and useful? Are the companies composed of truly comparable operations? If so, is data gathered the same way and organized in a comparable format? While these issues are important, so is the fact that much time can be wasted on measurement problems.

*Whom to benchmark, and who is the best?* Xerox regularly benchmarks all direct competitors, all their suppliers, and all major competitors to those suppliers. In addition, certain functions of renowned companies are benchmarked as potential performance standards. (L. L. Bean is a benchmark reference in mail-order distribution.)

Through Fuji-Xerox a benchmarking window is open into the Far East. The same is true in Europe through Rank-Xerox.

*Data collection.* This process generally starts with public sources of information: trade publications, annual reports, and open meetings. Occasionally consultants design and conduct surveys, but most of the time, data is directly exchanged with the benchmarked company. When comparing the details of internal operations, it's almost the only way. Benchmarking a circuit of companies can sometimes be done in return for sending data to each of them—if everyone understands that in advance. Quid pro quo benchmarking exchanges are starting to add to airline traffic. To be meaningful, data collection should be repeatable on a regular basis.

**2.** Analysis. Xerox asks four basic questions. (*a*) Is the competition better, and if so, by how much? (*b*) Why are they better (or worse)? (*c*) What can be learned? and (*d*) How can Xerox apply what was learned?

Updates are very important. Projecting how fast competitors are improving is as important as knowing their current capabilities.

Then determine the size of performance gaps, positive or negative, between Xerox and the benchmark company. The nature of the gaps, their magnitude, and a breakdown of reasons for them or components of them are very important. Try to develop an objective basis upon which to attack the gaps—to close negative ones or extend positive ones.

**3.** Integration. Xerox describes this phase in 15 steps that boil down to first *accepting* the results. A big negative gap is easier to accept if executives understand *why* it exists. Overcoming the denial syndrome follows. The benchmarking gap and general goals are widely communicated.

Strategies and action plans follow. For this process, Xerox relies on the problem solving of its employee involvement program.

**4.** Action. Any obvious corrective action is started immediately, but many solutions are not obvious. Most attacks are multidisciplinary. Standing teams such as commodity teams or new product teams swing into action. Special employee involvement teams are formed as necessary, and the employee involvement problem-solving processes move into high gear.

Tracking progress is essential. Benchmarks should be periodically remeasured to check goals, and progress against planned milestones mark the rate of internal response.

**5.** Maturity. The best form of maturity is to see that many benchmarking gaps are either zero or positive. While good progress has been achieved and Xerox is becoming the benchmark in certain areas, more often than not, Xerox finds that the elusive target continues to move. Another sign of maturity is to have benchmarking become a natural, integrated part of the ongoing process of continuous improvement, so that wherever Xerox does not happen to be the current leader, it is at least making a hard run at it.

Xerox has used most of the measures below for benchmarking. Each measurement is not always ideal in form, but it is usually possible to *roughly* compare different companies.

### Cost and Cost-Related Metrics
- Percent of cost of each function compared to revenue:
    Sales
    Service
    Customer administration
    Distribution
    General and administrative
- Labor overhead rate (percent).
- Material overhead (percent).
- Manpower performance ratio.
- Days, weeks, or months of supply.
- Cost per page of publication.
- Cost per order.
- Cost per engineering drawing.
- Occupancy cost as percent of revenue.
- Return on assets.

### Quality
- Percent of parts meeting requirements.
- Percent of finished-product quality improvement.

- Number of problem-free final products.
- Internal and external customer satisfaction rates.
- Billing error rates.

*Service*
- Work support ratio.
- $/SCAT hour (Standard Call Activity Time).
- Service response time.
- First time fix of service call problem.
- Percent of supplies delivered next day or on time.
- Percent of parts available for the technical representative.

Note: Improvement rates are easier to compare than absolute numbers.

Source: *Competitive Benchmarking: What Is It and What It Can Do for You*, by the Xerox Corporate Quality Office, Stamford, CT, 1984. (Called the "red book" within Xerox.)

and comparable companies. Looking at financial results alone does not reveal the competitive potential of a competitor. It is a situation similar to a coach evaluating scouting reports on a rival team whose record the previous season was mediocre. By knowing what to look for—size, speed, quickness, mental concentration on the game, play execution, and so on—a coach can tell if this year's opposing team will be formidable, regardless of their past record. However, in business as in sports, potential does not win; actual team performance at the crucial time does.

Benchmarking does not improve performance by itself. It does provide a means to prevent unjustified complacency.

## Comparison with Ultimate Standards

Comparison with a standard is a very familiar practice. Some standards are set based on the performance that planners believe a company needs to make in order to achieve its financial goals for the coming year. Budget plans lead to standard costs and variances. Their shortcoming is that they seldom are attached to measures that will motivate dramatic improvement, and the standards are usually expected to be "realistic," that is, not demanding breakthrough improvement.

Standards based on best-in-class operations elsewhere are better. Such standards may not be up-to-date when obtained, but many of them are certainly relevant. They represent moving targets. For example, no sooner does a PC board wave-solder operation hit an old BIC standard of 125 ppm (parts per million) defects than the company discovers that someone else has set the new BIC for comparable boards at 100 ppm or even below 50 ppm. Chasing world wide best-in-class standards is one way the phrase "world-class manufacturing" was coined.

Suppose a company is determined to be the best. Its targets of improvement must project beyond the currently known best-in-class. If the ambition is to be inspired to better performance by tough competitors, that approach is sufficient challenge.

The ultimate challenge is the ultimate standard. There are several examples: Zero defects, zero inventory, zero time lost for setups, zero trials of a new tool, and the value-added ratio. Many ultimate standards will not literally be accepted as serious goals of performance because humans cannot see themselves as perfect. There is no way to beat such a mark; one can only fall short of it.

People will accept targets that are close to perfection. Motorola's six-sigma quality goal is such a standard. It is not perfection, but it is close to a zero defect goal.

A zero inventory goal taken literally is impossible. Something must be present to work on. A close approximation is "one-piece flow," meaning that the only stock present is at workstations with value being added, and there is no accumulation between machines.

Ultimate standard goals, or close approximations to them, provide the insights and inspiration. One that is similar to the zero inventory goal, but more versatile, is the value-added ratio. It is most valuable where value-adding activity is easily defined. An example is the ratio of the time that value-adding activity is actually being performed on an automobile during assembly from the time that the main body panels are first spot welded together until completion.

Counting body, paint, chassis and final assembly operations, suppose that 1,200 vehicles are in the line and 1 is completed every minute. With 1,200 cars total on this line, there are 1,200 positions in which work can be done. However, work is not done in all of them; the cars are merely passing from one workstation to another. Unless an assembly plant has become conscious of the idleness of cars merely moving, it will not keep record of the proportion of time that value is really being added to the car body itself. (Note that this is not the same as worker efficiency or tool-use efficiency.) When such a record is kept, the value-added time is a low percentage of the total time cars are on the line—well under 50 percent. The value-added time ratio of parts at lineside is generally well under 1 percent.

By thinking this way, some old issues are revisited and new challenges are presented. For instance, quick-drying paint has been a quest of the auto industry since its beginning, but keeping auto bodies in sequence with minimal time for touch-up after painting is a fairly recent challenge to be taken seriously. Drying time is a paint technology problem. Touch-up-and-trim time after painting is consumed by many little problems amounting to inspection and rework. The value-added ratio forces some useful questions, such as how the idle time of bodies after painting could be shortened. This goal of reducing a value-added ratio translates into a shop floor discipline to question any action that permits a car body (or materials) to merely sit. That questioning generates ideas to promote flexibility and waste reduction.

## By Their Measurements Ye Shall Know Them

At the beginning of the 1990s many manufacturers seem to be stuck halfway between the business-as-usual paradigm of manufacturing and the "performance excellence" paradigm. They would like to adopt many practices of the new manufacturing, but be assured of financial benefits rather quickly.

The upshot is an inconsistent mix of old and new performance measurements.

A financially driven organization that has not begun the new manufacturing is usually circumscribed using numerous cost controls and financial ratios. For instance, inventory management performance is considered a separate function of the company, and it is measured by total inventory investment, inventory turnover ratios, and customer service from inventory.

As this same organization begins to streamline its operations, the measurement of inventory flow in inventory turns is likely to be restated as days-on-hand. A days-on-hand label emphasizes that inventory represents time and that to cut inventory we should cut wasteful practices that consume lead time.

However, for a long time after beginning a change to the new manufacturing, managements still need reassurance that results will show up where they have historically "really counted," in financial returns. Two of the measures that help organizations span the gap in thinking are inventory carrying costs and costs of quality (which are really costs of nonquality). Once aware how much waste is covered by inventory, companies increase inventory carrying-cost percentages with extras beyond the investment costs. Then inventory carrying cost reductions help to financially justify that a low inventory is better. Cost of poor quality performs the same role convincing managers that quality reform has a payoff.

Cost of quality is generally defined to have four components, the cost of: (1) prevention, (2) inspection, (3) internal quality failures, and (4) external quality failures. One way to convince the financially oriented that quality is important is by demonstrating that preventive quality practices reduce the overall cost of quality. In the absence of a cost system designed for flexibility, manipulations of source costs into a cost-of-quality format are sometimes heroic, but the cost of quality in the novice company is so high that the total quality processes will show a cost-of-quality reduction no matter how approximated.

Once a dramatic 30 or 40 percent cost-of-quality figure shrinks to under ten percent, the apparent potential for further cost savings is reduced, but by then management is much more convinced that pro-quality decisions are correct. Quality decisions are much less frequently subjected to cost trade-off decisions.

As companies progress in developing their quality processes, the scales of measurement change. For many, a fraction of a percent defect rate was good performance before total quality. With improvement, the scale for defects changes to parts per million. A half-of-a-percent defect rate "sounds" like good performance; a 5,000 parts-per-million (ppm) defect rate "sounds" as if it needed attention.

Not only do scales of measurement keep changing, more importantly the attention shifts from product defects to process quality—preventing defects. Instead of managers fretting about the percentage of defective product shipped, *workers* begin to worry about the percentage of operations that are not subject to a fail-safe methodology to catch defect-causing conditions before a defect is actually made. Performance measures shift from percentage of defective final product, to process capabilities, to percentage of operations covered by fail-safe methods, to the attention paid processes during unusual circumstances when the ordinary fail-safe methods are not functioning—as during start-ups or trials.

The same thing is true of other performance measures once considered the gold standard of production performance. Short-term efficiency measures disappear. Long-term total productivity measures start to be tracked. Shop-floor labor and materials tracking simplifies; in the repetitive manufacturing case, it nearly disappears. In its place emerges a system of visual control for production.

Would the new measures mean abandoning cost variances and efficiencies? Well, perhaps not totally. Much of the problem is the narrowness with which such measures have been interpreted, rather than the numbers themselves. Long-term improvement ought not be neglected to pursue short-term performance numbers.

Detailed, operation-by-operation variances and efficiency measures offer little guidance to long-term improvement and often mislead. They should be abandoned as wasteful. However, in a broader sense, a variance tells us that we missed a budget, and an efficiency is one measure of productivity. This information may be useful when interpreted in context of an overall improvement effort. Direction of improvement should be based on information from many sources: customers, physical processes, employees and others. Costs are an important component of this information, though only one type of measure. Needed: simple, easily interpreted systems for both cost and noncost measures of performance.

## NOTES

1. As recounted in E. F. Schumacher, "Toward an Appropriate Technology," *Atlantic*, April 1979, pp. 88–93.
2. Lawrence F. Gross, "Building on Success: Lowell Mill Construction and Its Results," *The Journal of Society for Industrial Archeology* 14, no. 2 (1988), pp. 23–24. (The author's main point is that from the beginning the mill owners followed "standard practice of creating adequate buildings at limited expense." Material handling was torture and machinery vibration a constant vexation. But no one could justify improvement.)
3. Noriaki Kano, "TQC in Japan and the U.S.," paper presented at GOAL/GPC Sixth Annual Conference, December 5, 1989.

# CHAPTER 3

## USING COST TARGETS TO CONTROL OPERATING PERFORMANCE*

### INTRODUCTION

In almost all American manufacturing companies, one of operating managers' main tasks in the past 40 years has been to manage costs reported by the accounting systems.[1] Traditional cost systems "roll down" cost targets from top-level planning budgets to lower-level operating reports. Cost targets are seen as an important tool to "control" the operating performance of plant managers and department supervisors. Like the setting on a thermostat that keeps a furnace working at the rate needed to maintain a desired room temperature, cost targets are viewed as a desired outcome to compare against actual performance. Variances between actual and desired results provide feedback that is supposed to prompt operating managers to adjust what they are doing.

Many authorities believe these traditional cost concepts and management accounting tools contributed to the decline of American manufacturing in the past two decades.[2] Indeed, one person has gone so far as to say "cost accounting is the number one enemy of productivity."[3] Companies on the road

---

* The material in this chapter may not be copied or otherwise used without the express written permission of H. Thomas Johnson or Arthur Andersen & Co., based on their copyrighted material from a future work tentatively entitled "Customer-Focused Activity Management."

to manufacturing excellence must ask if old cost systems and old management information hinder their efforts to achieve manufacturing excellence.

Companies are not likely to find operating performance information that supports manufacturing excellence simply by changing to a new set of accounting techniques. As the two preceding chapters make clear, the difference between "business as usual" and manufacturing excellence is more than one of technique. It is also a difference in management philosophy. Underlying manufacturing excellence is a people-oriented philosophy that is antithetical to the cost-oriented philosophy underlying business-as-usual cost accounting practice. Companies need cost techniques and management information that reflects this people-oriented philosophy.

In particular, operating performance information for manufacturing excellence views employees as an active force helping a company achieve its goals. As one well known airline says, "people make the company an on-time machine." However, cost-oriented business-as-usual companies train executives to manage "by the numbers." They define goals in terms of financial outcomes and tell people to achieve those goals by cutting costs or earning ROI. They don't deny people are necessary, but they see employees as passive cogs in a tightly controlled finance-driven system.

Traditional cost targets reinforce this passive role for employees. They tend to constrain, not enhance, people's actions and vision. For example, pressure to minimize standard cost variances often encourages department supervisors to keep machines and people busy producing output, regardless of market demand. By using such targets to encourage high utilization of worker and machine time, financially oriented companies often cause unnecessary inventories of finished and in-process merchandise to accumulate. They see product lead times increase, and their dependability at keeping schedules decrease. To achieve their separate cost targets, each department impairs the company's overall ability to compete. People left to their own devices produce better results. But people cannot move a company toward manufacturing excellence if they are rewarded for meeting independent, finance-driven targets: not in order to satisfy customers, internal and external.

Traditional cost-oriented performance measurement fails to support manufacturing excellence primarily because it motivates people to sustain output in order to achieve cost targets. This practice appears in many guises: going into overtime at the end of accounting periods to eliminate unabsorbed burden variances; striving always to use productive capacity to the fullest extent possible; or hiding excess output in one period to use as a buffer in another period. Managing output to control costs must stop as companies move toward manufacturing excellence.

The roots of this practice probably are found in the methods American manufacturers developed to cope with increased product variety in the last 50 years. Supplying variety to the market seems to be a necessary ingredient to long-run success in any business. However, variety breeds complexity. A long-run imperative for any business, then, is to minimize the complexity (and cost) of providing customers with a variety of products and services.

## DECOUPLED OPERATIONS, "EFFICIENCY," AND WORK

To reduce the complexity of managing product variety and to increase "efficiency," most manufacturers in the past 50 years maintained decoupled production operations so that each plant, office, or department could focus on one process in an operation (e.g., component assembly, invoicing, welding, order taking, heat treating). This decoupling and specialization affected all work in all types of manufacturing companies, from custom job shops to so-called continuous-flow process facilities.

Large, multinational, multiplant companies often subdivide an entire multiprocess network, of course, to reduce complexity (and its attendant costs).[4] They focus plants, for example, by coherent groups of product lines and by product age; or they focus full-line plants by geographic region. Focusing by subdividing an entire multiprocess chain of operations along such lines is considered an appropriate way to improve performance, because it reduces costs of complexity.

But focusing by decoupling operations into processes— depicted in Figure 3–1—has a different effect. While it can reduce local complexity and local costs in separate parts of a system, it may reduce the performance of the system as a whole. It does this by creating buffers and added work that increase system wide costs and impair a company's ability to respond adequately to customers.

Decoupling operations into separate processes creates a "flight of bumblebees" (see Figure 3–1, B) that increases buffers, space occupied, transportation, and time. It diminishes cooperation among departments and allows separate parts of the organization to work at independent rates, not all at the customer-demand rate. Over short periods, such as a month or less, rates among various parts of the system can, of course, vary considerably—with the differences going into or coming out of inventory. Theoretically the output from all parts of the system will balance out with customer demand over long periods of time. However, that balance is often forced by means such as building buffer stocks, scrapping excess output, reducing prices to clear out excess stock, and, in cases of severe imbalance, eliminating parts of the system by laying people off and selling assets.

These consequences of decoupling are exacerbated if cost measurements induce people to gain local advantages at the expense of systemwide results. Companies do that when they use traditional cost accounting targets to control "efficiency" in decoupled systems. Buffers and imbalance in decoupled operations have been promoted by most manufacturers in the past 50 years by two cost-minimization rules: (1) minimize *total* cost in any process stage by "optimizing within fixed production constraints" (i.e., trade-off constraints); (2) minimize cost of labor or machinery *per unit* of output (i.e., "be efficient") by utilizing workers and machines to the fullest extent possible.

Attempting to minimize costs by following the rules of optimization and utilization is a hallmark of cost-oriented business-as-usual. Managers manipulate output to optimize total costs and to minimize measured unit costs, not to supply what customers want. The ironic outcome of producing to manage cost instead of customer demand has been *higher* total costs as well as *impaired* ability to serve customers.

With growing variety and increasing process fragmentation, adherence to the optimization and utilization rules meant increased costs and impaired customer value for several reasons:

**1.** "Optimizing within setup time as given" caused managers to build faster machines with more capacity in order to accelerate output of pieces and subcomponents in every process of a production system. As variety increased, requiring more setting up, long setup times meant that less and less

**FIGURE 3–1**
**Decoupling Operations Adds Work**

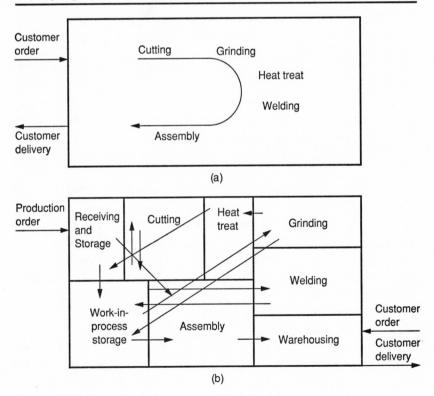

(a)

(b)

time on each shift was available for producing good units. Companies kept up with the demand for increased output, therefore, by pushing work out as fast as possible between setups, in the remaining time that pieces and subcomponents were actually being processed.

**2.** To recoup their cost, larger machines were "utilized efficiently" by scheduling work in longer, faster runs and larger lots. Faster rates and larger lots, increasingly in excess of the immediate customer-demand rate, meant increased inventories and storage.

**3.** Bigger and faster machines producing pieces and subcomponents for storage, not for immediate use, diminished the chance for feedback about defects. This created costly scrap and rework.

**4.** Allowing decoupled processes to work at disparate rates created a need for myriad support departments to transport material, to expedite orders, to schedule work and material flows, to monitor work and costs, and to inspect output.

Planning and scheduling complex inventory and material flows required complex systems, and the systems themselves added expense and time delay. Standard cost variance reporting systems deserve special attention because of the contribution they themselves make to cost and complexity.[5] Believing that total cost in the whole system is equal to the sum of the costs in each separate process, management accountants created standard cost variance systems to monitor costs in each and every process of a company's production system. For direct costs they devised labor and machine tracking schemes that reported direct costs per hour or per unit of output. For overhead costs they devised reporting schemes to track the percentage of overhead "covered" or "earned" by units produced.[6]

The goal of these reporting schemes is to have all recorded direct labor or machine hours go toward production of standard output and thereby "absorb" or "cover" direct and overhead costs—a condition referred to as "efficient." Intelligent department managers beat this system by scheduling workers and machines to produce output in long runs, so little time is needed for changeovers and setups—categories of indirect or "nonchargeable" time. They keep workers and machines busy producing output, because output enables a department to "earn" the direct hours incurred each reporting period. Every unit produced, including the equivalent of full units in partially finished work, entitles the department to a standard allotment of machine or labor hours. If a department produces enough equivalent finished output to "earn" all the direct hours reported in the period it is declared 100 percent efficient. It doesn't matter if the output is saleable. In fact, hours spent on "allowable" rework are often considered to be "efficiently covered." With so flawed a system, people sometimes put in hours creating defects, just to build inventory.

Therefore, achieving standard direct cost "efficiency" targets leads to larger batches, longer production runs, more scrap, and rework. Ironically, managers' efforts to achieve high, standard cost efficiency ratings have tended over time to increase a company's total costs and to impair competitiveness. In any plant striving to achieve manufacturing excellence, standard cost performance systems are anathema—especially if incentive compensation is geared to controlling standard-to-actual variances. Those systems encourage waste (excess production) and imbalance (end-of-period production spurts to reach "efficiency" targets). Moreover, they do not motivate continually changing and improving the methods and parameters of production.

But perhaps the most insidious consequence of standard cost performance evaluation systems is the insistence they place on managing *cost numbers* rather than managing *value* (i.e., competitiveness). They encourage managers to "control" costs by producing output to forecast, not to supply what customers want. Indeed, customers scarcely fit into the world of standard cost performance. The customer is merely someone the company persuades to buy the output that managers are driven to produce, at prices it is hoped exceed variable costs.

Companies that understand manufacturing excellence know these things. Harley-Davidson knows this. General Electric knows this. Both companies used traditional standard cost performance measurement systems years before they ever heard of JIT. When they began to implement JIT systems in their plants in the 1980s, both companies discovered that traditional performance measures *clash* with the imperatives of satisfying customers "just in time."[7]

The clash occurs as soon as plant personnel begin to spend their time to create value, not just to produce output. Creating value in a JIT system means taking extra effort to do things right the first time, pitching in wherever need arises, and training—all of which traditional labor reporting systems usually classify as "nonchargeable" time (i.e., time not "earned" producing output). Hence, doing what it takes to send customers the *right* output *on schedule* can reduce the department's efficiency rating. That's the clash.

## MEASURE TOTAL COST
## AND MOVE CONSTRAINTS

The answer, or at least a good part of it, at companies like Harley-Davidson and GE, is to evaluate shop-floor performance in terms of things that matter to customers—such as keeping promised delivery schedules and reducing calls to repair defects. However, they also monitor cost—*total, actual* product costs controlled by the entire plant, not standard unit costs of pieces and subcomponents tracked in separate departments.

Ironically, when plants in GE's Medical Systems Group first moved toward JIT in the mid-80s they discovered the corporate accounting system could not produce information about the total, actual product costs. Now those plants track *total* payroll (direct and indirect, not just direct payroll) and total material costs of products at the plant level. The company evaluates plant managers on their success at driving down total cost—an unambiguous index of something that adds value both to the customer and to the company, like defect rates and meeting delivery schedules.

Added complexity and cost caused by efforts to achieve departmental cost performance targets in virtually all industrial firms by the 1950s did not impair any one company's competitiveness and profitability since they all followed basically the same procedures. Most manufacturers by the 1950s used standard cost targets to "control" decoupled systems through which work lurched toward completion in discontinuous batches. The complexity of these systems caused added *time* (the time from iron ore to finished automobile went from $3\frac{1}{2}$ days at Ford's River Rouge plant in 1923 to weeks in most automobile plants by the 1960s) and *cost* (cost of people and resources consumed in storing, moving, inspecting, expediting, scheduling, monitoring, accounting, and waiting; not to mention the cost of investing in larger and faster equipment). But no one minded this added time and cost as long as all competitors did things the same way.

That condition changed after 1970. Some competitors, largely from Japan, discovered how to achieve variety without decoupling all the processes in a production system. The change began in automobiles, where companies such as Toyota recreated what Henry Ford had done at River Rouge in the 1920s, but this time with variety. How to achieve continuous flows with variety was not widely understood in the Western world until the early 1980s. Indeed, traditional cost performance systems have prevented many companies from understanding this change, much less taking the steps needed to make the change. But many companies have now discovered that the change entails simplifying all activity to achieve balanced, continuous flow of value-adding work.

Continuous simplification of the flow of activities is the excellent manufacturer's primary means of moving constraints, eliminating trade-offs, and thereby supplying more of everything customers expect at competitive prices. We already referred in Chapter 2 to the simplification that results from reducing setup time, not optimizing within setup time as a given constraint. Other examples of simplification aimed at moving constraints, not taking them for granted, include designing products to use common components and requiring vendors to deliver material and parts that meet statistical control tests. Both steps eliminate work and costs that otherwise seem necessary.

Simplification removes waste, which is manifest by delay, excess, and unevenness and caused by the practices companies adopt when they pursue "optimization" and "efficient resource utilization" in decoupled systems. Instead, the goal of simplification is to achieve rapid, continuous (i.e., coupled) flow of value-adding activity. Those were the characteristics of Henry Ford's River Rouge plant that produced one model in one color in the early 1920s. Those *also* are the characteristics of a modern customer-focused enterprise that produces any variety of models and colors the customer wants.

The route to simplification is paved by moving constraints, not optimizing within them. Eliminating constraints promotes simplification by enabling companies to continuously remove the work (and attendant time and cost) they add when their efforts to optimize costs in decoupled processes creates buffers and unbalanced flows. As our previous discussion of setup constraints demonstrated, efforts to optimize setup costs, say by producing larger sized lots, generates additional storage, transporting, reworking, scheduling, and other types of work.

That type of work is often referred to as "nonvalue activity." Certainly it gives customers no added satisfaction, even though it adds extra cost. But this work is necessary in getting the job done, as long as the constraints that cause it are not changed. Therefore, it is not really correct to say that companies achieve manufacturing excellence by eliminating nonvalue activity. More to the point, they eliminate the constraints that make nonvalue activity *necessary*. Once the constraints are removed, the work called "nonvalue" becomes unnecessary and can then be eliminated.

Changeover time is not the only constraint that makes nonvalue work necessary. Other constraints, and examples of the added work they spawn, include:

1. *Lead time* (magnified by long changeover times). Long lead times generate a demand for added scheduling, forecasting, and finished goods inventory.
2. *Plant layout.* Poor layouts generate a demand for transporting items, for more work-in-process inventory, and for additional setup time.

3. *Product design.* Poor product design generates a demand for parts ordering, work-in-process inventory, and rework.
4. *Process design.* Poor design of production processes and work flows generate a demand for inspection, rework, additional setup time, and unscheduled maintenance.

This list is not exhaustive. It simply indicates the scope for improvement processes. Enormous amounts of work, perhaps seventy percent or more of all work done in a business, can be eliminated by removing the underlying constraints that make it necessary.

The purpose of improvement processes is to eliminate constraints. By moving constraints, rather than optimizing within them, managers effectively eliminate trade-offs among sources of customer satisfaction such as quality, responsiveness, service, and price. Making nonvalue work unnecessary means that a one-time cost (usually nominal) of removing a constraint is not only offset by a permanent reduction in the cost of work eliminated. It is also offset by a permanent gain in customer satisfaction from shorter lead times, better quality, and quite often, lower price. Removing constraints instead of optimizing costs within fixed constraints can be an enormous source of productivity improvement in the factory, the office, and the corporate staff.

## IMPROVEMENT PROCESSES
## BECOME JADED

Decades of optimizing costs in decoupled systems has ingrained cultural attitudes in the minds of managers that block efforts to move constraints and simplify processes. For instance, many companies confront this block after they succeed at campaigns to reduce setup times in factories. After an intense but short period in which setup times are reduced by orders of magnitude, people congratulate themselves on eliminating much work-in-process and on meeting schedules more dependably. Then progress stops. An island of excellence in one plant or department is surrounded by an unchanged flow of waste in upstream and downstream activities. Sometimes, even the gains in lead time and inventory reduction in the factory slowly erode with the passage of time. What has happened?

The answer, frequently, is a failure of top management to abandon the cost-focused financial management mindset.[8] They view setup reduction as "just another tool to cut costs," not as one step in continuous productivity improvement achieved by continually moving constraints. They don't change the system of decoupled measurements and the policy of optimizing within constraints as given. "What you measure is what you get." Since we're getting the wrong results, we need new measures. Manufacturers need new measures to achieve competitive performance in the new environment.

## MEASURING IMPROVEMENT IN ACTIVITIES

Excellent manufacturers track success at achieving a rapid, continuous flow of value-adding activity. Measures of success exist at two levels. First are global measures of delay, excess, and unevenness such as the three ratios listed in the left-hand column of the following list.[9] These measures can be taken at any level in the organization, and they do not have to be taken frequently, perhaps only as often as once or twice a year. The right-hand column indicates the question asked by each ratio.

| | |
|---|---|
| Total time/value-added time | How much does the time spent to do something (e.g., prepare an invoice, assemble a product, or answer a customer inquiry) exceed the time one ought to spend if there were no delay? |

| | |
|---|---|
| Use rate/demand rate | How evenly are processes balanced to the final demand rate? |
| Number of pieces per workstation | How much does work-in-process exceed what is needed to exactly supply what the customer wants, *when* it is wanted? |

On a second, more detailed level, excellent manufacturers track progress in specific programs to create value or to eliminate waste, using measures posted prominently on charts in areas where the programs occur. These measures reflect trends—they are not standards against which to compute variances. The following list includes examples of such measures that might appear in factory workcells:

- Scores on housekeeping inspections.
- Setup times.
- Number of different items set up per shift.
- Unplanned equipment down time.
- Number of suppliers per input.
- Distances that work travels.
- Throughput time.
- Number of skills mastered per employee.
- Number of new ideas per employee.
- ppm defects on the line.
- First-inspection pass rates.
- Space occupied.
- Batch sizes.
- Customer satisfaction indexes.

Continuous improvement in the preceeding measures reflects elimination of waste and, therefore, improved customer satisfaction and reduced cost. However, managers who embark on journeys to manufacturing excellence must be aware that simplification itself does not *automatically* reduce costs. It automatically makes resources redundant; such as, it frees up time of workers who once moved or stored inventory and it releases space once used for aisles or storage. But wages or rent recorded in the books of account will not fall until people eliminate or redeploy such redundant resources. Managing redundant resources is a vital part of any company's efforts to achieve manufacturing excellence.

## MANAGING OPERATIONS TO ACHIEVE PROFITABILITY

Simplifying activities helps companies achieve manufacturing excellence. But simplification alone is not sufficient to guarantee profitable results. In the long run, unless costs of operations are in line with revenues that customers pay for the value they receive, a business will not achieve its financial performance goals. An enterprise can satisfy customer expectations and achieve competitive operations yet lose money if it does not control spending. Therefore, the enterprise needs reliable information to gauge the impact its operations have on long-run spending.

The information in traditional cost accounts is too highly aggregated to identify the impact of improvement programs on operating costs. In departmental operating reports that display variable costs, fixed costs, and usually some type of variance data, the costs reflect aggregate spending by object (i.e., people, materials, capital) and department. These reports show where dollars were spent, and on what, but say nothing about causes of the resource consumption that prompted the spending in the first place. For example, operating reports will show dollars spent in the maintenance department (*where*) for wages and supplies (*what*) but they will not show how activities in the purchasing department (acquiring low-priced but impure materials) cause spending on machine maintenance. To control costs, managers need more than traditional cost accounting information. They need information about the underlying causes and consequences of resource consumption.

Activity-based cost systems, the subject of the next chapter, may be a source for this information. Proponents of activity-based costing argue that costs are not caused by spending, the proximate source of information for costs recorded in accounting records, but by the consumption of resources in activities. Presumably, companies can not control performance by trying to manage accounting numbers, only by managing the activities that cause those numbers. Ideally, managers should know how reducing setup time, for example, affects the cost of resources no longer needed to store inventory, to occupy space, to transport work between processes, to inspect defects, to rework, to purchase oversized equipment, and so forth. Moreover, they should also know the financial impact on customer satisfaction of setting up more frequently in smaller lots. It is not yet clear, however, if today's activity-based costing systems are the first step in that direction.

Companies that practice activity-based costing must recognize that controlling operating performance and measuring financial results are two different things. To control operations they need information about measures of competitiveness—lead times, ability to meet deadlines, time to market with new products, and so forth—not measures of financial results. However, to estimate accurately the financial consequences of operations, companies will undoubtedly need information on the costs of activities. Activity-based cost information—information about costs of resources consumed in specific activities—enables companies to simulate the long-run financial impact of steps taken to simplify and improve operating activities. Activity-based costs can help companies assess accurately the costs of resources made redundant by simplification programs.

In companies that use activity-based costs, operating managers will not ignore accounting costs. On the contrary, on the wall of every department should be a chart showing "total costs down," as the Japanese call it. But people looking at such charts should always understand that what drives total costs down is adroit activity management, not time devoted to controlling costs. Any other accounting costs at the operating level simply distract managers from doing what is important. Cost information other than total costs down—including all forms of activity-based cost information—should be used primarily to evaluate strategic options, to plan, and to budget.

## LINKING PLAN AND CONTROL INFORMATION

Activity-based costs enable companies to link planning information in financial budgets with nonfinancial information used to control operating activities. The power to link financial planning information with measures of nonfinancial activity frees companies from the temptation to control operating personnel with financial information "rolled down" from planning budgets. American manufacturers have not always succumbed to that temptation. Indeed, before the 1920s it would have been rare to find such practice.

In fact, the company that virtually invented modern financial management—E. I. DuPont de Nemours Powder Company—seems not to have controlled operating managers with the financial information from its earliest ROI planning budgets.[10] In the decade before 1920 top managers at DuPont had detailed monthly statistics on the net income and ROI of every operating unit in the company (see Figure 2–1). But they seem never to have imposed net income or ROI targets on managers of their explosives manufacturing plants. Instead, plant managers followed targets dealing with timeliness of delivery to customers, product quality, plant safety, customer training (to use a very dangerous product), and comparative physical (not dollar) consumption of labor, material, and power among plants. Secure in their knowledge that plant managers would look after those key determinants of competitiveness, top managers took responsibility for the company's financial performance.

This division of responsibility disappeared in most American manufacturing companies after 1920, as the growing importance given to external financial reporting caused top managers to roll financial targets down into the operating levels. By the 1950s managers at all levels in almost any large manufacturing firm were pursuing financial goals, especially cost targets. It was in that context that companies decoupled processes and controlled costs by managing output—a strategy that worked reasonably well as long as every competitor did likewise. However, as we have seen, this strategy caused managers unwittingly to lose contact with operating activities. When manufacturers appeared in the 1970s who competed by organizing activities in a new way, practitioners of traditional financial management were caught off-guard.

Activity-based cost information eventually may allow top managers to compile financial budgets exclusively to plan and simulate—not to control operations. Top managers can "roll up" the expected costs of a company's operating activities to compile financial budgets, employing ROI targets or whatever, as they have done for decades. But now they do not have to roll those budgets back down into the organization. They can control operating performance with nonfinancial measures of customer satisfaction. At the same time, their activity cost information links nonfinancial information about operating activities with top-level financial targets. With activity-based costs, companies can simulate the financial consequences of operating plans while simultaneously controlling operating activities with measures that truly reflect value to customers (e.g., meeting schedule promises, defects at final test, lead times, and the like).

Even without activity-based cost systems, companies can nevertheless link nonfinancial measures of operating performance and top-level financial targets without having to roll down financial plan information to operating levels. Plants in GE, for example, are starting to do this by using the hierarchy of measurements depicted in Figure 3–2. As we mentioned before, some GE plants shifted operating-level measures from the old "earned direct labor hours" performance measures to newer measures of total cost and customer satisfaction, as shown at the bottom of Figure 3–2. These new measures were chosen with two important conditions in mind: (1) as operating-level measures improve, key success factors must improve; and (2) operating-level measures must improve

**FIGURE 3–2**
**The Hierarchy of Measurements at GE**

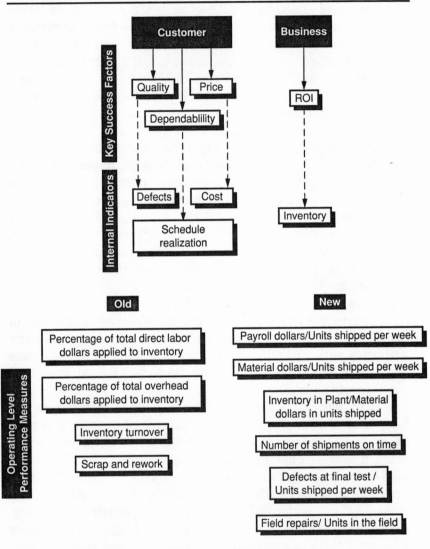

whenever the plant moves constraints and makes work redundant. Achieving those conditions avoids the problem with
traditional efficiency measures in which steps taken to improve efficiency or to control cost absorption can impair competitiveness and long-term profitability.

## CAPITAL BUDGETING

The subject of resource management concerns long-term as well as short-term outlays. Decisions to make long-term capital investments must consider, of course, the full cost of waiting for the future to unfold. That cost usually includes factors such as risk, time value of money, and expected inflation. Capital investment decisions must also capture the full range of customer values and resource commitments impounded in any future investment proposal. Its failure to capture fully the future impact of new technologies, and thereby inhibit continuous improvement and innovation, is perhaps the chief deficiency of the traditional financial approach to capital budgeting.

The traditional finance approach to capital budgeting usually assumes the future will be an extrapolation of the present. Financial investment analysis considers how new investments reduce current costs, but it usually assumes that current benefits will continue to flow whether or not new investment is made. A new investment is justified, then, largely by the present value of cost savings, primarily through the displacement of direct labor. This assumption may be valid where investment entails replacing old with new equipment embodying the same technology. But it overlooks the radically different benefits inherent in today's state-of-the-art technologies. This oversight biases against investments that offer improved quality, dependability, service, or flexibility.

Indeed, costing systems that stress direct-labor performance measures reinforce the tendency for American companies, in particular, to invest primarily in short-term, cost-reducing or labor-saving projects. Information from those systems, used in traditional capital budgeting techniques, tends to bias against any investment that promises to develop the company's people or to build customer satisfaction in the future without reducing current costs or direct-labor hours. To overcome this bias when evaluating investments in new technologies, companies must extend the time horizon and broaden the scope with which they view costs and returns.

However, companies probably should abandon traditional financial analysis altogether when evaluating R&D expenditures.[11] R&D is a special class of spending that often is mistakenly equated with investment spending. Investments presume knowledge, albeit imperfect, about future costs and returns. R&D represents an effort to gain a stake in future opportunities before anything is known about future costs and returns. Uncertainty about future outcomes has the opposite effect on current values of investments and R&D. Companies facing a future of new and often unexplored technological opportunities will impair their long-term financial well-being if they judge R&D with the same financial tools they use to evaluate investments.

R&D spending resembles the stake in the future one acquires by purchasing options in financial markets. Downstream benefits affect the value of an option quite differently than they affect the value of an investment. When evaluating an investment, downstream benefits are more heavily *discounted* the more uncertain and the more distant in time they are. However, the value of an option is greater the more uncertain the outcome and the longer the time horizon in which to capture the outcome. Uncertainty and timing of future outcomes can never increase an option's downside risk; that risk is limited to the amount spent up front to purchase the option. So also with R&D: companies that relate R&D spending to the certainty of a project's financial outcome are not giving due consideration to the danger of foregoing learning about new technologies.

To see how R&D resembles taking an option on the future, consider companies that began to explore automatic and electronically controlled machine tools when that technology first appeared in the mid-1970s.[12] Those companies were well positioned to exploit the microprocessor-based revolution in capabilities—much higher performance at much lower cost—that hit during the early 1980s. Because their operators, maintenance personnel, and process engineers were already comfortable with electronic technology, they found it was relatively simple to retrofit existing machines with powerful microelectronics. Companies that had earlier deferred spending

R&D funds on electronically controlled machine tools fell behind: they had acquired no option on these new process technologies.

For actual investment proposals, as distinct from R&D spending, it probably is sufficient to recognize that not all future costs and benefits of today's investments are captured in the money magnitudes customarily tracked by the double-entry accounts. Recognizing that fact, one can then go ahead and reckon the net present value of flows that can be reduced to such money variables. Then, if a project does not have the requisite net present value to warrant going ahead with it, one can consider how much net value would have to exist in the "imponderables" (such as deeper customer relationships, flexibility, etc.) to make it go. Usually managers will know if the unquantified variables are or are not capable of carrying the balance.

In fact, the "hard" numbers in traditional capital budgeting often rest on untenable assumptions that companies are less likely to adopt if they use people-oriented and activity-based performance information to evaluate investment proposals. Contrast, for example, how traditional capital budgeting and people-oriented budgeting might treat the cost of training in an investment proposal. Training would presumably have the same importance for the outcome of an investment project no matter how the proposal is evaluated. But the traditional approach, where short-run cost saving carries the day, might consider training expendable if it makes a difference to the net present value outcome. The newer management approach, however, where cutting lead time might be the primary goal of the investment, would identify training as critical to the outcome and not expendable regardless of cost.

## CONCLUSION

An important source of manufacturing competitiveness is product variety. However, complexity of operations stands in the way of achieving cost-effective variety. An ability to cost-effectively achieve product variety is an important source of competitive advantage. Removing constraints and simplifying activities open this pathway to competitiveness by attacking complexity of operations, thereby allowing companies to provide more variety at lower cost.

Eliminating constraints continuously eliminates the need for work that is caused when companies decouple operating processes, ostensibly to reduce complexity and achieve efficiency. Eliminating that work improves quality and increases flexibility and dependability. That is how simplification enables companies to eliminate the trade-off between fulfilling rising customer expectations and cost. Moreover, simplification enables companies to reverse "overhead creep," the insidious tendency for overhead's share of total cost to grow.

Manufacturers that achieve competitive excellence in today's economy look very different from those that operate according to principles of the old manufacturing paradigm. Here are two sets of paired terms that characterize differences between the old and the new manufacturing paradigms:

| *Old Paradigm* | *New Paradigm* |
|---|---|
| • Mass production | • Small lots |
| • Homogeneous output | • High variety |
| • Buffers | • Zero inventory |
| • Rework | • Right the first time |
| • Discontinuous flow | • Continuous flow |
| • Slow changeovers | • Fast setups |
| • High overhead cost | • Low overhead cost |
| • Long lead times | • Short lead times |

Achieving the characteristics in the right-hand column implies that a company is flexible, dependable, and capable of delivering value to customers. Indeed, the main implication of achieving manufacturing excellence is to come continually closer to a state where output is produced only to satisfy customer wants. A company achieves this state by building what Taiichi Ohno, "father" of the Toyota Production System, refers to as the two pillars of Toyota's system: produce exactly what the customer wants, when the customer wants it; and be able to see a defect and stop to correct it when it occurs.

Construction of these pillars is often impeded by performance measurement systems. Existing performance measures seldom suggest or support the steps companies must take to become competitive today. Indeed, traditional management accounting measures tend to give priority to internal demands of a manufacturer's own production routine, whereas the new manufacturing paradigm gives priority to development of people and satisfaction of customer demands. All companies travelling on the road to manufacturing excellence must resolve the clash between traditional performance measures and the people-oriented imperatives of the customer-driven global economy. Companies that resolve the clash will undoubtedly pass by those that do not.

A symptom of failure to resolve this clash is hearing a company's managers say it takes additional cost to provide more of anything people value, such as quality, service, flexibility, or dependability. That trade-off concept is deeply rooted in American management thinking. Moreover, traditional accounting performance measures motivate behavior and reinforce constraints that make trade-off virtually inevitable. To compete in the global economy, manufacturers must overhaul those long-standing performance measures. The payoff is enormous. By adopting the new manufacturing paradigm, companies continuously move constraints that pit cost against satisfied customers, both internal and external.

The key to manufacturing excellence is to remove constraints that cause trade-offs between cost and customer satisfaction. Traditional cost-oriented performance measures motivate behavior that makes trade-offs inevitable. Trade-off causing constraints are moved by simplifying activities and thereby eliminating waste.

World-class manufacturers recognize that long-run profitability is achieved by people creating value and people consuming resources. To control operating performance, therefore, they focus on activities people perform, not on costs of output. Moreover, they examine costs of activities to assess the financial consequences of improvement programs.

They also use activity-based costs to roll up estimates of long-term product costs, information useful for assessing the profitability of strategic product choices. The next chapter discusses this and other important uses of activity-based cost information.

## NOTES

1. The most comprehensive articulation of the cost concepts and tools mentioned in the first two paragraphs is found in the classic cost and managerial accounting textbooks authored by Robert N. Anthony and Charles T. Horngren.
2. For a comprehensive discussion of this viewpoint see H. Thomas Johnson and Robert S. Kaplan, *Relevance Lost: The Rise and Fall of Management Accounting* (Boston: Harvard Business School Press, 1987).
3. Eliyahu Goldratt, co-author (with Jeff Cox) of *The Goal.*
4. A theory of focused operations is set out in Wickham Skinner, *Manufacturing in the Corporate Strategy* (New York: John Wiley & Sons, 1978).
5. The evils of standard cost performance measurement systems are recounted both in H. Thomas Johnson, "Performance Measurement for Competitive Excellence," in *Measures for Manufacturing Excellence*, ed. Robert S. Kaplan (Boston: Harvard Business School Press, 1990), pp. 63–90, and in Robert S. Kaplan, "Accounting Lag: The Obsolescence of Cost Accounting Systems," in *The Uneasy Alliance: Managing the Productivity-Technology Dilemma*, ed. Kim B. Clark, Robert H. Hayes, and Christopher Lorenz (Boston: Harvard Business School Press, 1985), pp. 195–226.
6. A well-known description of these systems is in Charles T. Horngren and George Foster, *Cost Accounting: A Managerial Emphasis* (New York: Prentice Hall, 1988).
7. On GE, see H. Thomas Johnson, "Performance Measurement for Competitive Excellence." On Harley-Davidson, see William T. Turk, "Management Accounting Revitalized: The Harley-Davidson Experience," *Journal of Cost Management*, Winter 1990, pp. 28–39.
8. H. Thomas Johnson, "Performance Measurement for Competitive Excellence."

9. Adapted from Richard J. Schonberger, *World Class Manufacturing Casebook: Implementing JIT and TQC* (New York: Free Press, 1987), pp. xi - xxiii.

10. This discussion of DuPont Powder Company comes from Johnson and Kaplan, *Relevance Lost*, chap 4.

11. Graham R. Mitchell and William F. Hamilton, "Managing R&D as a Strategic Option," *Research: Technology and Management*, May–June 1988, pp. 15–22.

12. Robert S. Kaplan, "Must CIM Be Justified by Faith Alone?" *Harvard Business Review*, March–April 1986, pp. 87–95.

# CHAPTER 4

## ACTIVITY-BASED COSTING

A consistent theme of this book is the inappropriateness of using cost information to "run" a manufacturing plant. Symptoms of such a cost focus are an emphasis on achieving labor efficiency and increasing volume to absorb overhead even though the output is not required to meet customer requirements. Excessive inventory, poor quality, high cost, and low customer service are often the consequence.

We cannot dispense, however, with the need for cost information for strategic decision making. Strategic decisions are made, in part, on the basis of profit potential, and one side of the profit equation is cost. Decisions concerning what customers to serve, which products to sell, and what work to outsource all use cost information.

We have always assumed that cost was useful for strategic decisions. What we didn't realize was the extent to which conventional cost systems, especially systems that costed products, provided inaccurate estimates of cost. Many manufacturing managers, for example, can attest to believing that "product costs are way off." It has only been in recent years, however, that we have confirmed that their worst fears were correct.

The particular need for reasonably "accurate" product costs has stimulated the development of a new approach to costing. This new approach is called activity-based costing.[1] It differs from a conventional cost system in the precision with which it measures resource usage. In principle, such systems measure costs of any object, including products. In almost all cases reported to date, however, these systems have been implemented to cost products, usually with a major impact on reported product cost. It is common to find shifts in product cost that range in the hundreds of percent, and in some cases by thousands of percent.

Such shifts in cost are not academic because strategic decisions based on cost information affect profitability. The belief that certain products are profitable, for example, may lead a firm to expand its offering of those products. In reality, the firm may be better able to satisfy customer needs in other ways. Focusing on the wrong products will diminish its profitability.

---

**BOX 4–1**
**What Is Activity-Based Costing?**

Activity-based costing is an information system that maintains and processes data on a firm's activities and cost objects (e.g. products). It identifies the activities performed, it traces cost to these activities, and then uses various cost drivers to trace the cost of the activities to the cost objects. These cost drivers (such as the number of part numbers or the effort expended by product) reflect the consumption of activities by the cost objects. An activity-based cost system is used by management for a variety of purposes relating to both activities and cost objects.

Activity and cost driver concepts are at the heart of activity-based costing. Activities are processes or procedures that cause work and thereby consume resources. Examples of activities include tapping threads in a hole in a metal part; telephoning a vendor to place an order for materials; preparing a new drawing for a part to reflect a change in engineering specifications.

Cost drivers reflect the demands placed on activities by products or other cost objects. The cost driver "number of receipts," for example, measures the frequency of performing various activities connected with receiving and inspecting incoming components and updating the parts database. The cost driver "effort by vendor" measures the time required by various activities to establish and maintain vendor relations.

Cost drivers measure the demands placed on activities at both the activity and cost object levels. For example, the product cost driver "number of production runs" may show that the production scheduling department schedules 4,000 batches of product over a year. It may also show that Product A is run 6 times in a year and will therefore be scheduled 6 times.

Cost drivers are also used to trace cost to objects other than products. These objects include customers, markets, distribution channels and engineering projects. The activity "maintaining customer relations" is a customer sustaining activity and should be traced to customers via a cost driver such as "effort by customer."

Activity-based cost systems vary in the information provided on activities. Some systems use multiple cost drivers and allocate pools of cost to each cost driver. Each cost pool contains the resources consumed by several activities, but the activities are not separately defined. These "cost driver" systems are economical ways of reporting accurate product cost and are useful for strategic and product design purposes. In the absence of detailed activity information, as described in Chapter 3, they do not directly support the management of these activities.

## WHY CONVENTIONAL SYSTEMS FAIL

The major flaw in conventional product costing systems is their inability to report "accurate" product costs in most manufacturing settings. This is because conventional systems rely exclusively on drivers that are attributes of the product unit. Unit-level drivers include direct labor hours or dollars, machine hours, material dollars, and units of production.

Unit-level drivers can accurately trace the cost of activities that are performed on the product unit itself. Direct labor hours or machine hours, for example, are used to trace the cost of the drilling of a hole in a metal part to that part.

Unit-level drivers do not work, however, if there are activities not applicable to the unit level. Setting up a machine, for example, occurs at the batch level and benefits all units in the batch. Making an engineering change to a product occurs at the product level and benefits all units of that product.

Reported product cost will be inaccurate in the presence of such non-unit activities when volume or product diversity exists. Volume diversity is the presence of high volume and low volume products (or components) in the same manufacturing facility. Product diversity is the presence of products that create different demands for activities.

Table 4–1 illustrates volume diversity. Products A and B each require the same number of direct labor hours and so pick up an equal amount of engineering cost. This means that product A, the high volume product, receives the lion's share of engineering cost because there are more units to which cost is attached. This is not consistent with reality, however, because A and B each receive equal attention from engineering.

Table 4–2 illustrates product diversity. Product A is a mature product that requires little engineering attention. Product B is a new product and requires the resolution of several design issues and should receive more engineering cost. The direct labor hours driver, however, traces more cost to product A than to B because A requires more labor hours.

The impact of such errors on reported product cost can be substantial. Firms that have installed activity-based systems find that the cost of high-volume standard products typically

**TABLE 4–1**
**Volume Diversity**

|  | Product A | Product B | Total |
|---|---|---|---|
| Production volume | 1,000 | 100 |  |
| Number of engineering changes | 2 | 2 | 4 |
| Cost per engineering change | $1,100.00 | $1,100.00 | $4,400.00 |
| Direct labor hours per unit | 2 | 2 | 2,200 |
| Engineering change cost per direct labor hour ($4,400/2,200) |  |  | $2.00 |
| Conventional overhead cost per unit ($2.00 × 2) | $4.00 | $4.00 |  |
| Total conventional overhead cost | $4,000.00 | $400.00 | $4,400.00 |
| Activity-based overhead cost per unit [A = ($1,100 × 2)/ 1,000 B = ($1,100 × 2)/ 100] | $2.20 | $22.00 |  |
| Total activity-based overhead cost | $2,200.00 | $2,200.00 | $4,400.00 |

Source: ABC Technologies, Inc.

**TABLE 4–2**
**Product Diversity**

| | Product A | Product B | Total |
|---|---|---|---|
| Production volume | 1,000 | 1,000 | |
| Number of engineering changes | 2 | 10 | |
| Cost per engineering change | $1,000.00 | $1,000.00 | $12,000.00 |
| Direct labor hours per unit | 3 | 2 | 5,000 |
| Engineering change cost per direct labor hour ($12,000/5000) | | | $2.40 |
| Conventional overhead cost per unit (A = $2.40 × 3 B = $2.40 × 2) | $7.20 | $4.80 | |
| Total conventional overhead cost | $7,200.00 | $4,800.00 | $12,000.00 |
| Activity-based overhead cost per unit [A = ($1,000 × 2)/ 1,000 B = ($1,100 × 10)/ 1,000] | $2.00 | $10.00 | |
| Total activity-based overhead cost | $2,000.00 | $10,000.00 | $12,000.00 |

Source: ABC Technologies, Inc.

go down 10 to 30 percent, whereas the cost of low-volume specialty products may increase from 100 to 500 percent. Figure 4–1 shows the ratio of activity-based product costs to conventional costs for one company. The curve that results is typical of companies that have installed activity-based systems.

**FIGURE 4–1**
**Ratio of Activity-Based Product Cost to Conventional Cost**

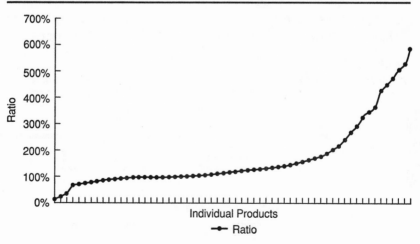

Individual Products
Ratio

It is discomforting for managers who use the inaccurate product costs to learn that decisions based on these costs were most likely wrong. If these managers had received accurate product cost information, would they have manufactured and sold the same products? Would the prices of their products have been the same? Would they have designed and produced products in the same way? In company after company we find that the answer to these questions is *no*.

It is not surprising that conventional costing systems report inaccurate product costs. Products make "cross-functional" slices across manufacturing organizations (Figure 4–2). It is unlikely that they would consume engineering, procurement, production, marketing, order entry, and distribution activities in the same ratio as they consume a single activity such as direct labor.

Conventional cost systems not only systematically distort product cost, they send signals that encourage actions that are inconsistent with manufacturing improvement programs. A cost system that shows that low volume, specialty, and expedited products cost no more than standard, high-volume products, for example, may encourage marketing to add or change the requirements for these products more than would normally be desirable. This is particularly true where the marketing function is physically separate from production forcing it to rely on the cost system to provide feedback on the impact of its actions.

This was exactly the case at the Wake Forest plant of Schrader Bellows. This plant produced over two thousand

**FIGURE 4–2**
**Products Are Cross-Functional Slices; Functions Are Vertical "Silos"**

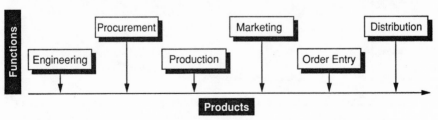

Source: This figure is based on a chart prepared by Alistair Duncan, of World Class International, Ltd.

products, representing about twenty thousand components, in batch sizes that varied from one to tens of thousands. Schrader committed resources toward many activities supporting the production of batches of product, such as setup, first-piece inspection, and material movement. The conventional, cost system failed to attribute cost to batches of product and therefore reported product costs that were markedly different from the ones reported by the activity-based cost system.[2]

The wide variety of products manufactured was, in part, a result of marketing's response to the reported cost of the products. The conventional cost system failed to reflect the true costs of diversity. Low-volume specialty products, for example, appeared to cost the same as high-volume, standard products. Marketing responded to this information by adding large numbers of low-volume specialty products to the product mix. Many of these variations were unnecessary to meet customer needs, and served only to increase complexity.

The loading of overhead cost via direct labor sends a signal to engineering and production that direct labor is very expensive. This is particularly true if a declining labor base is loaded with a large overhead pool and the overhead rate looks like a Latin American inflation rate of the 1980s. Engineering and production are encouraged to make changes that reduce the direct labor content of products or attempt to make the use of direct labor "more efficient." In reality, waste may increase and customer value may diminish.

Engineering may modify the design of an existing product to reduce its labor content and associated, reported overhead. This may be accomplished, for example, by replacing high-volume, hand-inserted components with new machine-insertable components. The activities required to support the hand-inserted components are already in place and no additional resources are required. The new machine-inserted components, however, require additional effort including new part number support, vendor relations, and purchasing. It is possible, therefore, that the reported cost of the re-engineered products goes down while overall overhead and cost go up.

Production may react to direct labor-based systems in several ways. Automation may be introduced incorrectly to try to reduce direct labor cost. Production may study direct labor in excruciating detail and build complex and costly systems to monitor its performance and improve its efficiency. Batch sizes may be increased and volume emphasized in a misguided attempt to "increase efficiencies" and to absorb overhead via additional direct labor hours. As was shown in the previous chapter, however, these actions can increase overall cost and may reduce quality and flexibility.

Another flaw in conventional cost systems is their functional orientation. Costs are accumulated by line item, such as salaries, and then by function, such as engineering, within each line item. Reports for each functional manager are prepared listing all the line items of expenditure for that function. In some cases, these reports show the budgeted, actual, and variance amounts for each line item.

This functional orientation does not fit with the cross-functional reality of modern manufacturing. The engineering-change process, for example, requires activities in several functions (Figure 4–3). The engineering change initiated in product engineering, for example, requires changing the production procedures in manufacturing engineering, updating the production schedule in production control, phasing out obsolete components and procuring new components in materials management, and retraining personnel and revising procedures in the service department.

What is lost by accounting for resource consumption by line item and function? First, the multiple activities that exist in most functions are not visible. For example, the inspection

**FIGURE 4–3**
**Engineering-Change Processes Will Cross Functional Boundaries**

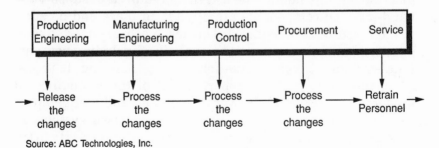

Source: ABC Technologies, Inc.

function may include first-piece inspection, finished-goods inspection, and complaints inspection. A review of the department budget, however, will reveal nothing more than salaries, supplies, and other line items of expenditure; this information is of little benefit in the management of the activities.

Second, conventional cost systems encourage a functional rather than a cross-functional orientation. Functional classifications of cost provide only partial information on activities that cross functional boundaries. The engineering change process described in Figure 4–3, for example, is several activities linked across functions. No single function "sees" all the activity required to successfully execute an engineering change. No single function "owns" the process.

Third, functional classifications of cost encourage the use of financial performance measures, such as cost variances, that reinforce a functional orientation. These measures are typically heavily emphasized and are the source of behavior that improves functional performance at the expense of overall performance. The use of the purchase price variance to evaluate the purchasing department, for example, encourages changing suppliers to reduce purchase cost, but may also reduce quality and disrupt production.

Conventional cost systems, therefore, hinder manufacturing improvement programs in several ways. Inaccurate product costs encourage marketing decisions that increase waste and complexity in production. An exclusive focus on unit-level drivers such as direct labor hours encourages behavior that interferes with design for manufacturability and continuous improvement programs. The classification of costs by line item and function fails to support the management of activities, drivers, and cross-functional processes and inhibits cross-functional cooperation.

## WHY ACTIVITY-BASED COST SYSTEMS SUCCEED

Activity-based cost systems cope with volume and product diversity and therefore report more accurate product costs than conventional systems. This is because activity-based cost systems recognize non-unit-level cost drivers, such as the number of engineering change orders, in addition to the unit-level drivers of conventional systems.

If we return to Table 4–1, we can see how activity-based cost systems cope with volume diversity. The objective is to pick a cost driver that measures each product's demand for the engineering-change activity. Because each engineering change requires the same resources, the number of engineering changes per product meets this objective. (In practice, engineering changes vary in the amount of effort required. A different cost driver, such as estimates of the number of engineering hours per product, is required to reflect these differences.)[3]

The number of engineering changes traces an equal amount of engineering cost to each product in Table 4–1. This is because each product consumes an equal quantity of the driver (two changes each). The cost per unit is lower for Product A, however, because the volume is lower. Both these calculations are correct because each product receives equal engineering attention but the volume of units that benefit varies.

The cost driver "number of engineering changes" also copes with product diversity in Table 4–2. It traces substantially more engineering-change cost to Product B in this example because B is a heavy consumer of the engineering-change activity. This translates into a higher engineering-change cost per unit for B because its volume is the same as A.

The improvement in the accuracy of product cost that results from using activity-based costing has a major impact on marketing strategy. Managers naturally chase profits, and prefer to focus their efforts on profitable products, product lines, markets, customers, and distribution channels. Profitability for each of these objects is a key strategic performance measure for many companies. It is to the *information* in the cost system, however, that managers turn to assess their profitability.

The introduction of an activity-based cost system is an opportunity to review the impact on long-term profitability

**BOX 4–2**
**A Glossary of Activity-Based Costing**

*Activity*—an activity is a process or procedure that causes work and thereby consumes resources. An example of an activity is entering the details of a customer order at a computer terminal.

*Unit-level activity*—a unit-level activity, such as tapping threads in a metal elbow, is performed on a unit of the product.

*Batch-level activity*—a batch-level activity, such as scheduling the production of a batch of parts, is performed on a batch of a product.

*Product-level activity*—a product-level activity benefits all units of a product. An example is maintaining the routing sheets or bill of materials for a product.

*Facility-level activity*—a facility-level activity, such as janitorial work, benefits the process as a whole rather than an individual product.

*Cost driver*—a cost driver is a measure of the frequency and intensity of the demands placed on activities by cost objects (see *cost object*). An example is the number of part numbers.

*Activity center*—an activity center is a cluster of activities that are related by function or process. The engineering change process, for example, may be reported as an activity center.

*Cost object*—a cost object is an entity that requires the performance of an activity. Cost objects include products, customers, markets, distribution channels, and projects.

*Bill of Activities*—the bill of activities is a listing of the activities and associated cost required by a product (or other cost object). The bill lists activities in various levels of detail or simply the cost drivers required by the cost object.

of decisions affecting product mix, pricing, customer support, and other strategic issues. Such decisions should be made in the context of the ability to eliminate waste and thereby reduce the cost of manufacturing products and supporting customers. It may not make sense, for example, to drop unprofitable, low-volume products if setup times can be reduced significantly.

A manufacturer of circuit boards, for example, often received orders for one or two replacement boards from its customer service division. These boards were costed—and priced—at the same level as the boards in production at much higher volumes. Production of replacement boards, however, often involved a special engineering effort and special handling throughout the plant. When the company installed an activity-based cost system it confirmed their belief that these low-volume boards cost many times the standard cost. They responded by taking steps to make the production of replacement boards no more costly than production boards. They also repriced the replacement boards at more realistic levels.

The circuit board manufacturer example highlights the two options for responding to the insights derived from an activity-based cost system. One option is to change product strategy and the other is to increase strategic capability via manufacturing improvement. Changing product strategy is the easier of the two options. The firm can give up markets and reprice products without making any change in operations. Reducing diversity in this way improves competitive position if the firm is more focused on customer needs. For example, it may eliminate unnecessary product variation that adds only complexity to the task of manufacturing without providing additional value to customers.

Reducing diversity, however, is a second best strategy for many firms. Dropping products and walking away from markets runs the risk of "cutting off one's tail one piece at a time." The remaining products and markets may remain uncompetitive (and are the next candidates for the axe) because no steps have been taken to eliminate waste, improve quality and customer service.

In contrast, improving strategic capability via manufacturing improvement increases the ability of the firm to profitably meet diverse customers' needs. Reducing setup times or eliminating activities associated with receiving, inspecting, stocking, and moving parts, for example, reduces overall cost. It also reduces the reported cost of low-volume specialty products. This creates an opportunity to meet the unique needs of individual customers.

Activity-based costing is therefore an important strategic tool. It reports costs that are more accurate than conventional costing systems. More accurate costs of products, for instance, reduce the likelihood that prices will be set incorrectly or that unprofitable products will be sold aggressively. There is a danger, however, that a firm will respond to the insights derived from activity-based costing by reducing diversity rather than reducing the cost of diversity via improvement. Reducing the cost of diversity increases the strategic capability of the firm by allowing it to profitably meet varied customer requirements.

## IS ACTIVITY-BASED COSTING MORE THAN A STRATEGIC SYSTEM?

Originally developed as a product cost system to be used primarily for strategic purposes, activity-based costing is used today by some firms to support improvement programs. These uses include evaluating the impact on cost of alternative product designs, confirming the reduction in the cost of activities and products that result from waste elimination programs, and better understanding the flow of work in cross-functional processes.

These new uses take advantage of the different type of information found in activity-based cost systems. Conventional product cost systems use unit-level drivers that encourage attention to unit-level efficiencies and absorption, which work against efforts to improve manufacturing. In contrast, activity-based cost systems identify the activities required by each cost object and the driver quantity for each activity. For costing products, they use batch and product-level drivers in addition to unit-level drivers.

It is useful to compile a bill of activities containing the information on the activities and drivers required by each cost object. If the product is the cost object, the bill of activities is a road map for each product's cross-functional journey to the customer. It can be shown in various levels of detail. The bill of activities in Table 4–3, for example, lists the activities and associated cost required by a product. A summary version of the bill would show the cost drivers required by the product but not the activities associated with each cost driver.

The bill of activities can be the starting point for cost-reduction efforts. The bill of activities is a listing of the inputs required by the design engineers to produce and distribute the product they have designed. The influence of product design over the bill is so great, some companies believe that 70 percent or more of the cost of a product over its life cycle is locked in upon completion of the design. Operating people often tell us that their most important cost driver is product design. They point to the number of process steps or activities relating to quality, such as testing, that are largely beyond their control.

The activities and drivers in the bill *are*, however, under the control of engineering! A group of engineers at Tektronix, for example, made the reduction of cost of the overhead activities a major objective in the design of a new generation of product. Overhead cost represented 30 percent of the manufacturing cost of the old generation of the product, with material cost and a small amount of direct labor the remaining 70 percent. The overhead cost of the new product was 9 percent of manufacturing cost.

**TABLE 4–3**
**Bill of Activities**

| Activity | Cost |
|---|---|
| Purchasing raw material | $ 1.87 |
| Receiving raw material | 2.19 |
| Forming molds | 2.94 |
| Moving molds | 1.24 |
| Pouring molds | 4.27 |
| Shaking molds | 1.76 |
| Annealing | 0.93 |
| Shearing and grinding | 1.10 |
| Hauling parts | 0.73 |
| (Other) | 7.29 |
| Total product overhead cost | $24.32 |

The product engineers at Hewlett-Packard's Roseville Network Division built their own activity-based cost system to allow them to incorporate cost information into their design decisions. Their objective was to design for manufacturability. This was crucial at Roseville because of the large number of products, their short life cycle, the continuous introduction of new products, and their production in lot sizes of one. The engineers had to pay careful attention to the design of the products to reduce their overall manufacturing cost.[4]

Engineers at Hewlett-Packard use activity-based costing to pick the least expensive activities for a product given equivalent functionality. For example, the engineers can select a component that requires either axial or DIP insertion. Because the cost system uses the number of axial and DIP insertions as cost drivers, the engineers are able to estimate the impact on cost of these alternatives. Table 4–4 provides a complete list of the cost drivers used in the Roseville system.

The engineers at Hewlett-Packard may not have made the right choice between these alternatives without the information from the activity-based cost system. Their intuition told them, for example, that axial insertion was a lot cheaper than DIP insertion. It required the activity-based system, however, to confirm their intuition and to calculate the relative cost of the two activities.

Product design determines a product's activity requirements, but there is usually considerable leeway in determining how the activity is carried out. Managing activities—by reducing the time or effort required—is at the heart of manu-

**TABLE 4–4**
**Cost Drivers at Hewlett-Packard's**
**Roseville Network Division**

| | |
|---|---|
| Number of axial insertions | Number of radial insertions |
| Number of DIP insertions | Number of manual insertions |
| Number of test hours | Number of solder joints |
| Number of boards | Number of parts |
| Number of slots | |

facturing improvement programs. For this, one does not need cost information other than to gauge the cost consequences of decisions to change activities.

For example, large reductions in the time and effort required to setup a machine are possible without using cost information. An existing machine setup requires three to four hours. Proper tool placement, the development of standard procedures, performing tasks off the line, training and practice ("Grand Prix Pit Stop Teams") can reduce this time to a "single minute." The setup can also be performed by the operator rather than a specialist as a by-product of the standardization and training.

An activity-based cost system, however, can be used to gauge the cost consequences of such improvements. At a large electronics company, for example, the activity-based cost system showed that material-handling cost drivers, such as the number of receipts, the number of movements, and the number of inspections, accounted for a substantial portion of overhead cost. These activities occurred at the component level, and their cost often exceeded the material cost of a component by orders of magnitude.

These revelations were the stimulus to a waste reduction program. An examination of the current approach showed why it was so wasteful—it required a lot of people to perform the activities. The purchasing department used the computerized shop-floor scheduling system to determine what needed to be purchased. On-hand quantities in the stockroom were verified and phone calls placed to vendors. Purchase orders were prepared and mailed. Once the parts were received they were inspected, counted, moved, stored, and paid for.

Further study showed that thousands of low cost "C" components accounted for the bulk of this activity. A number of alternatives for these "C" components were developed and modeled using the activity-based cost system. One alternative was to order all low-cost components once a year. A second alternative was to order the "C" components twice a year. A third alternative was to source all "C" components from a single vendor, issue a single purchase order that would cover an entire year, supply the vendor with a three-month production schedule, and have the vendor replenish *kanban* bins weekly. The third alternative not only changed the frequency of work, but also substantially changed the way it was carried out.

The activity-based cost system was used to simulate the cost of each of these alternatives. The cost of each one was computed by estimating its driver requirements and multiplying the driver quantity by the cost-per-driver unit.

The analysis showed that the third alternative, using the single vendor and *kanban*s, was the low-cost alternative by a substantial margin. The potential cost savings were computed based on processing one purchase order and 12 invoices per year (versus thousands under the current system). The savings also resulted from the complete elimination of receiving, counting, unloading, inspecting, moving, and stocking activities for a whole class of parts. In addition, it was also observed that this alternative provided increased flexibility, reduced lead times and lessened the likelihood of component obsolescence.

Ideally, such improvements may be achieved *without* using information from an activity-based cost system. In the case just described, however, the ability to calculate the cost impact of improvements focused management's attention on the waste in material-handling activities. Activity-based costing also played a role in convincing skeptical managers that the changes were valuable because it permitted their economic impact to be estimated.

It is possible that activity-based costing fills an important need of Western managers for an analytical tool to support manufacturing improvement programs. Japanese-style, continuous improvement programs are often characterized by a "religious zeal." Managers who subscribe to these programs need "faith" or "intuition" that they work. Intuition may not be enough, however, to convert the nonbelievers, particularly when the nonbelievers are upper management. Activity-based costing fills this void by revealing the extent of current waste and demonstrating the impact of proposals to eliminate waste.

Using a cost system to estimate the cost impact of improvements does not ensure that cost will go down. Reducing cost requires two steps that, first, change the way work is performed, and, second, redeploy the resources that are freed up by the change. For example, the number of activities and time required to process a customer order may be reduced. This elimination of waste in the order-entry process permits the transfer of personnel to other activities. Overall cost is only reduced when this redeployment occurs.

Activities must be managed in the context of a cross-functional process that involves a supplier and an internal customer for each activity. The successful completion of an engineering change in Figure 4–3, for example, requires the linking of activities in several functional areas. Successful completion of the new design specifications in product engineering will not by itself ensure the timely production of a quality product.

Activity-based costing can support the management of processes by providing information that clusters activities and shows how they link together. Clusters of activities are activity centers. The activity center in Figure 4–4 shows several activities relating to the material-supply process. It is a reporting device that brings together in one place all the activities, and their associated costs and drivers, related to a process.

Some activity-based cost systems also provide process flow information that shows the flow of work. Figure 4–4 shows the flow of work from one activity to another and from one activity center to another. This overlay of process-flow information on the activity information creates a powerful tool for managing activities that goes far beyond the capabilities of early activity-based cost systems.[5]

One large manufacturing enterprise, for example, created activity centers to highlight processes that cut across functional lines. Activities relating to poor quality, for example, were performed in various departments. The "poor quality activity center" provided an understanding of the overall economic impact of activities required to detect and correct poor quality (the "cost of quality"). While the activity-based

**FIGURE 4–4**
**Material Supply Activity Center with Process Flow Overlay**

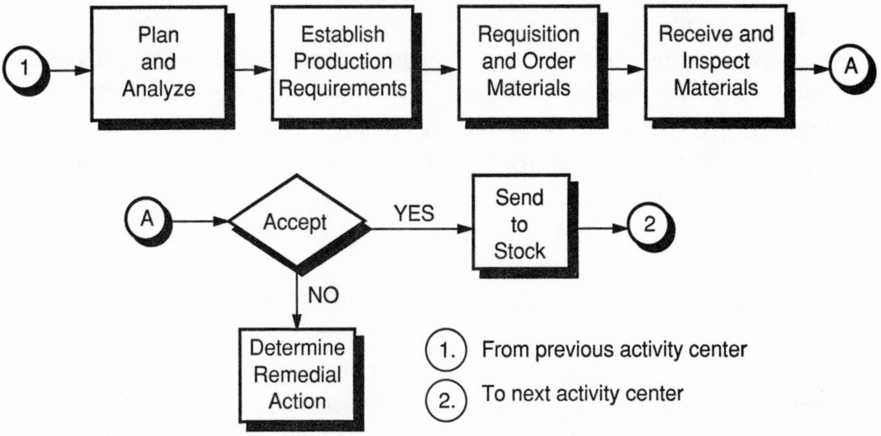

Source: ABC Technologies, Inc. This diagram was used as a diagnostic tool in a very complex environment (the main product had over 50,000 components, and there were over 200 overhead departments). The plant had an extensive infrastructure, multiple overhead departments, and processes that cut across department lines.

cost system did nothing to change this company's approach to quality, it did allow "ownership" of the process to be assigned, and encouraged the management of the activities within it.

## IS ACTIVITY-BASED COSTING JUST ANOTHER EXPENSIVE FINANCIAL SYSTEM?

Plants that have advanced along the path toward manufacturing excellence have in some cases devoted energy to turning off complex financial systems such as labor performance measurement, inventory tracking, and variance reporting systems. They see these systems encouraging behavior, such as building for inventory, that is inconsistent with manufacturing excellence. These systems also require inordinate amounts of time on the part of manufacturing and accounting personnel.[6]

Does activity-based costing take us back along the path toward unnecessarily complex systems? After all, activity-based systems require considerable operational data on activities and drivers. If the "number of receipts" is selected as a driver, for example, it is necessary to determine the number of receipts for each component.

Experience has shown, however, that activity-based cost systems are quite economical and unobtrusive. Much of the data required by activity-based costing already exists in the plant or is easily captured. The number of receipts per component, for example, is often stored in a computerized procurement system. All that was required was to download the data to an activity-based cost system.

Tracing cost to activities requires estimates of the effort expended by personnel on each activity. This effort is readily estimated via observation, examination of documents, or interviews. This indirect measurement provides results that are sufficiently accurate for the purposes of activity-based costing. It is not necessary to set up elaborate time-tracking systems to obtain more precise measurements.

Some people may not resist the desire to build large transaction-based reporting systems incorporating activity-based costing. They believe this may have the desirable consequence of making the *official* financial system consistent with the activity-based system and avoiding conflicting signals. Even if it is possible to do, it may have the undesirable consequence of making the system costly and unresponsive to the needs of its customers.

Existing activity-based cost systems today are typically small, stand-alone systems built on personal computers. They are versatile analytical systems serving the information needs of engineering, production, and marketing. They are under local control and are updated only when required, such as once a year.

The term "financial system" is not an adequate description of activity-based costing, because an activity-based cost system is strategic, not financial in nature. It is often the manufacturing or engineering organizations that bring activity-based costing into the company and not accounting or finance. In many cases, it is manufacturing that introduces activity-based costing to the company. At Hewlett-Packard's Roseville Network Division it was the design engineers who built the first activity-based cost system.[7]

Activity-based cost systems are typically built by cross-functional teams. These teams include individuals from engineering, production, and marketing in addition to accounting. This participation testifies to the value they attribute to activity-based costing and its cross-functional applications.

It is likely, therefore, that manufacturing companies of the 1990s will have multiple, "personal," activity-based cost systems in addition to plantwide models. These low-cost systems will support individuals in various functions as well as cross-functional teams.

## IS A SIMPLE SYSTEM THE MOST APPROPRIATE ACTIVITY-BASED COST SYSTEM FOR MANUFACTURING EXCELLENCE?

A simple activity-based cost system is one that has few drivers and recognizes few activities. Such a "frugal" system is the system of choice when manufacturing is simple—a focused factory (or a few similar products), few activities, and low overhead cost. A more complex system would represent waste. A simple system also permits focus on key aspects of manufacturing improvement programs such as reducing the part count or reducing lead times.

One should take care, however, to be sure that this simple system encourages the right behavior and reports accurate product cost. The Portable Instrument Division of Tektronix, for example, implemented a system with two drivers; the number of part numbers and direct labor hours. The purpose of the driver "number of part numbers" was to focus the attention of the design engineers on the cost of part number proliferation. The engineers responded by increasing parts commonality in new products and reducing the division's part count.[8]

While this initial response was appropriate, it was clear that commonality was only one of several important engineering objectives. Reducing the part count did not necessarily reduce the number of process steps required nor increase the quality of the product. The division expanded the number of drivers in the cost system to supply the engineers with the more detailed information they required.

A simple system may also report inaccurate product costs and lack credibility. Zytec, a manufacturer of power supplies, used throughput (cycle) time and supplier lead time in their new cost system. Throughput time, the elapsed time from the arrival of raw material in the plant to the shipment of the product, was used to trace manufacturing overhead to the products. Supplier lead time, the elapsed time from placing an order for a component to the time it was delivered, was used to trace material overhead to the products. The objective was to make the cost system consistent with the company's mission. Throughput time reduction was intended to reduce cost and improve quality and service. Supplier lead time reduction was intended to reduce cost and improve flexibility.

The system failed, however, because it did not report accurate product costs. Both drivers measure elapsed time, yet components and products consume activities in ways not reflected by time. For example, a vendor with a long lead time may deliver high-quality components, whereas a vendor with a short lead time may deliver low-quality components. A product with a long throughput time may require little engineering attention, while a product with a short throughput time may require substantial engineering time.

The response to Zytec's system was quite negative. The throughput time portion of the system confused management, and they were unable to explain the differences in cost from one product to another. Management also had difficulty understanding the relationship between supplier lead time and cost. Customers were unwilling to rely on prices based upon costs reported by the new system.[9]

The lessons are clear; simple cost systems work well in simple manufacturing settings only if used with care. Simple systems provide the benefit of low cost of design, implementation, and operation. A simple system is also easy to understand and focuses attention on key manufacturing performance measures. However, the system must focus attention on the right behavior or else it will be counterproductive. It must also report accurate product costs or else it will lack credibility.

## CONCLUSION

Activity-based costing differs from conventional costing in several ways. Activity-based costing traces cost to cost objects using different types of cost drivers. In systems that cost products, these cost drivers include batch-level drivers such as the number of production orders, product-level drivers such as the number of vendors, and unit-level drivers such as the number of insertions. Conventional costing, in contrast, uses only unit-level drivers. Some activity-based cost systems also provide information on activities, while conventional systems provide information on functions and objects of expenditure.

Conventional cost systems provide barriers to the improvement of manufacturing. Inaccurate product costs can adversely influence manufacturing strategy. The exclusive use of unit-level drivers such as direct labor hours encourages behavior that hinders design for manufacturability and continuous improvement programs. The classification of cost by function and object is of no value in the management of activities, drivers, and cross-functional processes. It may also prevent cross-functional cooperation.

Activity-based cost systems support manufacturing in ways that conventional systems fail to do. Strategic decisions benefit from accurate product costs. The availability of information on activities and drivers permits the calculation of estimates of the cost of alternative product designs. It also permits the calculation of the cost savings from actions to eliminate waste.

If designed and used properly, activity-based cost systems support manufacturing excellence. They can be economical to use, do not usually require additional data collection systems, and may be under the control of operations or engineering. Finally, simple activity-based cost systems work well in simple manufacturing settings if designed with care.

## NOTES

1. The discussion of activity-based costing in this chapter draws on the seminal work in the field by Robin Cooper of Harvard Graduate School of Business Administration. Cooper articulated the basic theory of activity-based costing in a four-part series of articles published in the following issues of *The Journal of Cost Management*: Summer 1988, pp. 45–54; Fall 1988, pp. 41–48; Winter 1989, pp. 34–46; and Spring 1989, pp. 34–46. This chapter also makes use of Professor Cooper's work on the four-stage hierarchy of cost drivers, i.e., unit level, batch level, product level, and facility level. Cooper first articulated this cost-driver hierarchy in a 1988 Harvard Business School working paper (revised through April 1989 under the title "Unit-Based versus Activity-Based Cost Systems") which in its final form will be published in the Fall 1990 issue of *The Journal of Cost Management*.
2. Robin Cooper, "Schrader Bellows," 9-186-272 (Boston: Harvard Business School), 1986.
3. The choice of cost driver is also influenced by the cost of measurement and the amount of cost involved. Engineering hours, for example, may be more costly and intrusive to track than the number of engineering changes. Also, the cost of engineering changes may not be high enough to justify the additional effort.
4. Robin Cooper and Peter B. B. Turney, "Hewlett-Packard: The Roseville Network Division," 9-188-117 (Boston: Harvard Business School), 1989.
5. Peter B. B. Turney, "Activity-Based Costing: A Tool for Manufacturing Excellence," *Target* 5, 2 (Summer 1989), pp. 13–19; H. Thomas Johnson, "Managing Costs: An Outmoded Philosophy," *Manufacturing Engineering*, May 1989, pp. 42–46.
6. See, for example, William T. Turk, "Management Accounting Revitalized: The Harley-Davidson Experience," *Journal of Cost Management* 3, 4 (Winter 1990), pp. 28–39.

**7.** R. Cooper and P. Turney, "Hewlett-Packard: The Roseville Network Division."

**8.** Peter B. B. Turney and Bruce Anderson, "Accounting for Continuous Improvement," *Sloan Management Review* 30, 2 (Winter 1989), pp. 37–48.

**9.** Robin Cooper and Peter B. B. Turney, "Zytec Corporation (B)," 9-190-117 (Boston: Harvard Business School), 1990.

# CHAPTER 5

# ORGANIZING TO MANAGE
# BY FACTS

The goals of the new manufacturing are accomplished through people. To reach the goals, an improvement process must take place through the people of the company. There are many diverse reasons for this improvement process. The most basic reason is to develop all employees in all parts of the company until they apply a scientific method, such as PDCA, throughout all operations.

The goals of the old manufacturing all too often were diverted to only making money. Even when the goals were broader, the concept of many managers was to translate the ideas of the brilliant few to customers through the directed efforts of the many. Only professionals diagnosed and solved problems of importance, and many companies never reviewed the rigor with which their professionals approached problems.

Measurement is part of the scientific method. Measurement helps identify problems, which are defined as performance falling short of that desired. Workers as well as managers should set targets as goals of improvement. Actual results are measured to see if goals have been met, or if corrective action was effective.

How measures are used is just as important as what is measured. In manufacturing, many performance measures relate to the personal performance of people, so the measurement process is not free of emotion. The types of measures used and how they are used, just as the approach to problem solving, depends on how a company is organized.

Organizations are changing. To carry out the new manufacturing, tall hierarchical forms of organization are evolving into something else. Companies are in various stages of reorganization. Most larger ones are, at a minimum, engaged in removing layers of middle management to create a flatter organization. Some, like XEL, are radically changing their shop organization to replace supervisory systems with self-directed work teams.[1]

## THE REASONS FOR FUNCTIONAL SILOS

Large industrial organizations have typically had problems integrating their various functional specialities. The symptoms of this problem are numerous: large committees, memos sent to a page full of addresses, long indecisive meetings, intricate action-approval procedures, and other practices known to corporate denizens simply as "bureaucracy."

The basic cause of much bureaucracy is strong separation between functional departments, sometimes referred to as "walls between departments." One of the more catching phrases is *functional silo syndrome*.[2] This analogy likens corporate systems of tall functional hierarchies to farm silos, each remote from the other, communication difficult among them, and filled with specialized materials. A person can spend an entire career in a functional silo without having any clear idea what goes on outside it. Suffering from the functional silo syndrome, separate departments of the same company can come to believe that their specialty alone is the major part of the company, and that their real competition is not with other, complete companies, but different departments of the same company, all competing for the same corporate recognition and resources.

There are many antecedents for hierarchal organizations in industry. Hierarchies have existed in almost every other form of human organization. One can come to believe that huge, rigid hierarchies are inevitable—basic to the human condition. However, large bureaucracies are the antithesis of simplicity and flexibility. Most larger manufacturers are working to circumvent the problems of functional separation in some way.

A major reason for functional hierarchies in manufacturing is that specialized problem solving is easier—as far as it goes. One person could cover a narrow scope of experience in greater depth. Technical competence as both a medical doctor and a metallurgist, for instance, has become nearly impossible, and this technological diversification has led to segregation in scientific and technical training. Professionals of different backgrounds have precious little in common in problem recognition and solution.

During the recent clamor over the state of public education in the United States, it is easy to overlook that many factory workers were never *hired* to use communication and problem-solving skills. The educated, developed worker was an underused resource. Even the contributions of the skilled-trade worker were often limited. Now that workers are expected to participate in improvement, we are finding that their personal development has stopped well short of their potential.

The professions in manufacturing, as we know them, began taking shape late in the 19th century. Engineering and accounting emerged as two of the initial functional professions. They have been splintering into ever-finer divisions since. In fact, the movement to recognize new specialties, such as certified purchasing manager, is still continuing.

The difficulty of integrating deep specializations has long been recognized. In a 1930 essay, "On the Barbarism of 'Specialization'," the late Spanish philosopher José Ortega y Gasset noted that it was possible to make scientific progress by dividing science into small sections, enclosing oneself in one of these to the exclusion of others, and generating voluminous results without a rigorous overall framework of thought that might make such work a rational advance.[3] Naturally, such "scientific barbarity" filled the philosopher with disdain.

Ortega y Gasset continued that this approach to discovering new knowledge also led to strange types of men—specialists—knowing very well their own tiny corner of the universe, but remaining largely ignorant of the rest. Sometimes they are even disturbed when confronted with an alien system of thought from another specialty.

Ortega y Gasset apparently had little experience with large industrial firms, but his essay has a strong counterpart in the sociology of the manufacturing firm. Without a strong common purpose and a common problem-solving methodology (goals and an improvement process), the various specialties fragment as they choose. The goal of profitability is not sufficient to bind them together. A more emotional change is necessary than redrawing organization charts. Changes in philosophy and attitude must take place.

Measurement systems play a role perpetuating functional silos. People become accustomed to measurements, rewards, and status systems based on their functional specialty. Arguments concern which department is responsible for what, and by what measurements its performance should be judged. A favorite is finished goods, for instance. No matter what department is appointed to "own" finished goods, other departments influence the size of the inventory and the size of write-down losses. Similar controversies surround "controllable and noncontrollable" aspects of other operating measures such as on-time delivery and premium freight costs.

There are many other disadvantages of functionally separated organizations. Communication is difficult—the memo-blizzard-and-endless-meetings problem. Work uses only a fraction of the talent range of many people stuck in narrowly defined jobs. Sequential decision making is slow. Information wends from one department's action queue to the next. Engineering changes are a typical example.

Response to customers is awkward when no one department deals with the full range of a customer's problem. Electronic communication per se is no solution. Perhaps it merely allows the handling of more complex issues in the same old way.

With all these disadvantages, why put up with functional organization? Because it is natural. Splitting by functions is easier. Provided the right problem comes to the right person's attention, addressing a detailed problem in depth takes less effort for a specialist accustomed to it. The case for functional organization is so familiar it is not always articulated.

Functional organization is large-scale organization. Fumbling cross-functional details and delaying decisions are acceptable so long as there is great value in *big* enterprise. As the competitive needs of manufacturing turn more and more to speed and quality in pursuing market niches, the value of large-scale functional organization is decreasing. However, the performance measurement systems of large-scale organizations live on.

## THE INTERNAL CUSTOMER

One of the favorite ways to start cracking the walls of the functional organization is by promoting the concept of the internal customer: the next operation is my customer. In a "JIT factory," the idea of the next operation as a customer is the organizational extension of so-called pull systems of production control. The basic idea is to give the next operation, and the person identified with it, exactly what they need when they want it, with no delays and no marginal quality. Part of designing a JIT factory is arranging layouts so that supplier operations can easily see customer operations. If that is impossible, "suppliers" should at least have easy access to their "customers."

Even in a factory where standard flow paths of material are not possible, the ideal is for each worker to know who will use the output from his or her work. The idea carries over into management and staff work—perhaps *especially* into management and staff work. The concept of the internal customer is integral to the improvement process as practiced in many companies. The short version of a six-step improvement process incorporating the internal customer is:

1. Identify customer.
2. Define the customer's specifications.
   (Define how to measure.)

3. Define the existing process.
   (Often involves measurement.)
4. Propose change.
5. Follow up.
   (Measure results.)
6. Standardize.
   (Involves measurement.)

This process requires a very different form of thinking from that in a functional hierarchy, and so does the measurement. Many companies are involved with this kind of improvement measurement, while at the same time judging departmental operations as a whole largely by old measurements.

Managements can easily overlook how different the measurement systems are. If the corporate accounting and finance departments are left outside the logical flow of the organizational shift to internal customers, they can subject the same people to two different measurement systems with different logics. In most companies the concept of the internal customer is still just an overlay on a more conventional organization; but under pressure, there is a big difference between working to improve by measures indicating the satisfaction of internal customers and working to improve by measures of performance considered important by functional bosses, or in particular, by remote and uninvolved corporate departments who still want to see performance in the usual mode.

Suppose a company making roller bearings decides to reduce waste by tightening production variances with little or no capital spending. Assembly is the customer of both race grinding and roller grinding. Because of the dimensional variances, races and rollers are both sorted into matching dimensional classes for assembly, a common practice in the bearing business.

If the grinding of races reduced variance much more quickly than the grinding of rollers, the assembly operation could be left without the extreme sizes of race needed to use the extreme sizes in the distribution from roller grinding.

The materials efficiency at assembly might go down until the improvement process at roller grinding catches up.

Under pressure for short-term performance, the assembly department and the overall production manager could easily decide to take the easy way out: to continue sorting a broadly dimensioned distribution of races and rollers and forget the total improvement process even though the end result would be higher quality bearings to the customer.

In this case, the cost variances are easily explainable, once reviewers understand the nature and direction of the improvement process, and that is the point. There are many ways conventional efficiency and variance measurements can go askew while people are in pursuit of improvement. Here are a few "classic" instances:

- Manufacturing overhead is underabsorbed when production schedules are cut to decrease inventory.
- Obsolete inventory paradoxically increases when parts are left over after redesigning a product line to increase the commonality of parts.
- If the introduction of preventive maintenance shifts maintenance tasks from maintenance personnel to machine operators, the initial effect is that the efficiency of both direct labor and maintenance labor decreases.
- If overall material handling work decreases, but the part done by direct labor increases while the overhead rate on direct labor remains unchanged, the result is an apparent increase in production costs.

In virtually all such cases, the problem of understanding is less in the measurements themselves than in how they are interpreted. The assumption of many such measures is that line workers and managers have the duty merely to operate systems and processes that have been well-designed by specialized corporate functionaries. Therefore any operating measure should detect whether designed processes are at all times being operated to generate profit according to plan.

But the existence of improvement processes dedicated to both design and operations improvement goals, often keying on internal as well as external customers, changes most of the assumptions of business-as-usual measurement.

## COMPANY STAKEHOLDERS

The term *stakeholder* became popular in manufacturing America in the 1980s. It plays on the word *stockholder*. Companies have many groups of people with interests in them in addition to the owners, and stakeholder is a term for those people.

When a large company is about to go under, as Chrysler was in 1980, the stakeholders in a company take visible interest in its survival. The stakeholders instrumental in Chrysler's survival were stock owners and creditors; employees, both union and management; suppliers, dealers, and customers who had contracts; and the part of the public, which saw itself economically linked to the company. In the end, even the federal government was a stakeholder. Chrysler was both a government contractor and a source of tax revenue.

The functional silos inside large companies are paralleled by the stakeholder silos associated with the company. Unless survival is uncertain, the stakeholders likely have different views of the company. If the occupants of each stakeholder silo are unable to appreciate viewpoints from the others, the same kind of confusion and conflict takes place among them as between the functional silos.

Legally and historically owners are the dominant stakeholders. The company is their possession, rightfully acquired by having put money into it, and those who risk their money have the say over what the enterprise does. As the late inventor, Bill Lear, once said about participatory management, "You put up half the money, and you can make half the decisions."

The assumption behind ownership dominance is that those with money at risk will most likely (but not assuredly) behave responsibly toward the other stakeholders, and that if other stakeholders dominate they will more likely serve their own interest at the expense of the rest. In this age of lever-aged buyouts, the assumption of enlightened ownership still depends on whether those on the control end of the leverage are short-term traders or long-term stayers.

In any case, anemic quarterly earnings by a publicly-held company can result in a stock price sag, which in turn invites potential raiders to start buying shares. Even a management vitally concerned with stakeholder fairness and long-term issues finds itself up against cold reality. If it yields mediocre earnings in the short term, then it will not be around for the long term.

In the computer era, financial results can be reported quarterly. In many publicly-held companies, the books are closed monthly. Greater investment in manufacturing companies is by fund managers. They too, may favor a long-term outlook personally, but they cannot afford it professionally: if underperforming the market in quarterly results, then investors pull their money from the fund. In much of the investment world, long-term survival depends on short-term performance—or those under pressure feel that it does.

Top managers are obligated, as the agents of ownership, to operate a company on the owners' behalf, but boards turn this into a short-term obligation with incentive plans based on quarterly earnings and stock prices. No wonder that performance measurement portrayed by the financial cost system retains great strength. Without question, the influence of this performance measurement system supersedes all others.

The insidious effect of this measurement system permeates companies and demoralizes personnel several stages removed from short-term investment. At the end of financial reporting periods, at quarterly or even monthly closings of the books, long-term waste fighting and quality practices are set aside in order to achieve the forecasted earnings and unit cost numbers. So much for continuous improvement processes.[4]

Perhaps the problems and conflicts in performance measurement systems are only symptoms. Causes lie elsewhere.

Consider one cause to be narrowness of stakeholders' interests as well as the myopic view from inside corporate functional silos. Table 5–1 shows the self-interests of only three stakeholder groups and how they might affect the emphasis placed on performance measurement.

Perhaps another cause is that the new manufacturing is based on people. Mass production belonged to those who assembled the capital to finance it, but competitiveness in quality, dependability, flexibility, and innovation depends on the development of people skills—interactive people skills in addition to the deepening and broadening of individual technical talents. The kinds of problem-solving teams that must deal with bringing products to market within supershort lead times will cross several stakeholder boundaries, not just departmental boundaries. One term for this is "supply-chain management."

It is hard to undertake long-term development on short-term money. The people-development mentality and the trader mentality mix like oil and water, but many managers are bravely trying to reconcile the two.

Accomplishments necessary to be competitive in the new manufacturing are measured in a time frame different from those necessary in short-term investment. In the new manufacturing, operational lead times may be dramatically decreasing, but investing in the human development necessary

**TABLE 5–1**
**Stakeholder Interests and Measurements**

| Stakeholder | Regards Company as: | Stakeholder's Goals from Co. | Stakeholders's Measurement Interest |
|---|---|---|---|
| Owners | Investment | Return on investment | Financial results |
| Customers | Supplier/ Servant | Quality, service, value | Quality, service, value |
| Suppliers | Customer | Profit, continued business | Customer satisfaction and future needs |

Goals and interests are interpreted here from the vantage of stakeholder self-interest just to make the point. Owners and investors want to know much more than the bottom line, of course, and every stakeholder associated with a company has some interest in financial results.

to execute quickly is more like investing in wine (not known as a JIT business).

Interactions of manufacturing personnel across company boundaries, and the broadening of skills necessary to perform in groups, signify that manufacturing is operationally becoming more like a service business, such as architectural firms or advertising agencies. However, the outlook of the investment community has not caught up with this development. We need to promote a form of customer-service capitalism for long-term investors in the new manufacturing.

## DEFEATING HYDRA-HEADED MEASUREMENTS

Strong separation of stakeholder interests and functional responsibilities in a company feeds a hydra-headed monster of performance measurement. Each separate group has a different viewpoint on what is important in measuring performance. Worse, a collection of independent measurements of different aspects of performance does nothing to promote teamwork. Different faces of the monster come from different directions.

In most functionally organized companies, the situation is not as stark as portrayed in Table 5–2, but everyone familiar with functionally organized manufacturers can think of numerous personal examples of stress caused by different performance measures pulling in different directions. Many of these measures also concentrate attention on the short term. There is evidence that short-term, routine reward structures stunt the inclination of people to engage in problem-solving behavior. [5]

Managements spend a great deal of time and ingenuity concocting performance measures that will lead to each function performing expertly in its own area, but at the same time cooperating with other functions. This battle with the hydra-headed monster takes place in meetings arguing the merits of various methods for sharing responsibility across functions. A

**TABLE 5–2**
**Examples of Stereotyped Functional Measures**

| Function | Responsibilities | Measures of Performance |
| --- | --- | --- |
| Finance | Earnings | Earnings to forecast |
| Accounting | Budget | Variances to budget |
| Sales | Sales | Sales volume to quotas |
| Materials | Inventory levels | Inventory turns |
| Purchasing | Purchased cost | Purchases price variances |
| Customer Service | Delivery, returns | On-time delivery, complaints |
| Production | Production costs | Labor efficiency |
| | | Equipment utilization |
| Quality | Process quality | Warranty rates |
| | | Reject rates |
| Design | Product quality | New product cost |
| | and performance | Product performance |
| Manufacturing | Machine design, routings | Cost per unit |
| Engineering | | ROI on equipment |
| Industrial | Labor standards, | Cost per unit |
| Engineering | productivity | |
| Maintenance | Equipment upkeep, repair | Equipment availability |
| | | Maintenance costs |
| Personnel— | Employee complaints | Overall labor rates |
| Industrial Relations | Training | |
| | Pay system | |
| | Employee classifications | |
| | Employee activities | |
| | Fringe benefits | |

few classic examples: Should a department be a cost center, or a profit center that must "sell" its services internally? Who should "own" finished goods inventory? Who is responsible for forecasting, and therefore for forecast accuracy? Should the value of customer-returned goods be allocated among different departments?

Measuring individuals within groups presents a similar set of dilemmas. Most companies rate individuals on some basis to determine their merit pay. For a good many raters as well as ratees, the process is as pleasant as a root canal. There are apt to be more disappointing outcomes than pleasant ones, and an individual who has been lumped into an average category for several periods in a row may cease to have motivation for improvement. By any forced-rank rating system, about half of us are below average.[6] Furthermore, everyone's performance is affected by the performance of others far and near.

Following his famous adage that an individual controls only 15 percent of his performance results and the "system" 85 percent, Dr. Deming advocates abandoning individual performance measures that peg individuals in performance categories for reasons beyond their control. However, most people by contention try to influence factors beyond their immediate control. For example, sales representatives vie for assignment to lucrative territories. Machine operators on incentive plans establish seniority rights to jobs with standards easy to beat. An ambitious manager assigned to a department, such as maintenance, where a brush with catastrophe is the only attention-getter, immediately seeks to amend the performance measurement system so that any form of improvement will demonstrate a magnified impact. The hydra-headed monster lives in a bottomless pit of motivational quagmires.

Expecting people to entirely give up making comparisons with each other is probably asking too much. Performance comparisons begin very early in American life. Ask any parent of quarrelsome preschool siblings. Later, even if a little league coach has a policy of playing and developing all players, carping will come from both the players and the sidelines. The hydra-headed monster can be defeated, but not slain.

This measurement monster threatens with its multiple heads: misplaced individualism, functional silo organizations, narrowness of view, and status systems built on making a set of management numbers. The performance measurement system helps perpetuate this state of affairs because a person cloistered in a functional silo and governed by its limited performance criteria can at first understand continuous improvement only in a very limited way. Typical comments include: "I see, but what's JIT going to do to my labor efficiency?" "I can't add to my freight cost just to make Materials look good." "We're already into TQC. The quality department makes sure no defects get to our line."

Companywide teamwork for improvement is inhibited by fixed, compartmented performance measures. The improvement process itself, and the measures that are necessary for it, must prevail over the hydra-headed monster.

Subjugating the monster is part of continuous improvement. Like a caricature of the devil, the monster lurks everywhere in organizational measurement, short-circuiting improvement processes before they can begin. For example, by "tradition," switching to a lower-price supplier that also meets quality and delivery standards is hailed as a victory—a positive purchase-price variance. However, the *limited scope* of this measurement never suggests: checking whether the new supplier will accept quality feedback, or has an improvement process in-place to continue improvement; checking whether the switch broke up a family of parts placed with one supplier; checking whether the *total cost* in the long run will be high.

Concentrating on a measurement such as purchase-price variance while neglecting other considerations is a natural extension of measuring functional performance with a limited number of measures. It forces decisions based on limited knowledge. It forces people to disclaim responsibility for total processes that no one really understands. ("I don't know why the new parts don't work as well as the old parts. They all passed qualification. Ask Quality Control.")

Engaging in problem-solving improvement between customers and suppliers, internal or external, is quite different from attaining a few numbers to please a boss. The major break with a hierarchal system comes when the customer's assessment of performance dominates. Something beyond mere profit-center logic is necessary. Workers and managers must seek how to please their customers, defining customer satisfaction and measuring it in as much detail as necessary. Customer measures of good performance, once customers themselves must investigate their own processes, cover several aspects of performance.

Taken to full development, this organizational approach makes as radical a transformation in purchasing, for example, as in any other part of a company. Some commodities may be purchased. Quality-critical items are provided through a problem-solving network formed with suppliers. Life as a model customer within this network is much more diverse than existence in a purchasing hierarchy where career progression seems to depend on only one measurement that management cares deeply about—purchase-price variance. Vision broadens. Roles change. Measurements are part of a mutual improvement process rather than an instrument of top-down control.

In a problem-solving network, the measurement system should contribute to corrective action with a minimum of assigning and shifting blame. In the ideal sense, every employee should become a manager, even when their job is designated as being a technical specialist.[7]

## THE PROBLEM-SOLVING NETWORK

Americans are now groping for a new philosophy of organization to replace the mechanically structured organization in manufacturing, or at least to ameliorate the effects of the functional organization. Changing the form of organization is an old subject. American companies have been through project management and matrix management as structural ways to overcome the shortcomings of functional organization.

One of the most popular ways to circumvent functional organization in the new manufacturing is an extension of matrix management by new names: task forces, project teams, or simply, "teams." In most companies it has proved difficult to develop worker teams to go beyond straightforward shop-practice changes. Cross-functional teams step into this void, spanning the gaps between functional groups and carrying improvement further than a worker-composed team can be expected to do.

However, a growing sense is that we have only begun to explore a new form of organization. A matrix of overlays on a rigidly structured organization becomes too complex. Western organizations must change the relationships between people. The new wave of management literature and "isms" are in search of a new definition of what work is and how people must relate together to accomplish it.

Little consensus is emerging from the management literature beyond generalizing that the new form of organization will be loosely structured. The titles of several books suggest a continuously shifting non-structure: *Thriving on Chaos, Dynamic Manufacturing,* and *The Change Masters* are three of them.[8] These three books have different thrusts, but they agree that people need empowerment and that organizations must be structured (or unstructured) to learn quickly.

The Japanese see that the new manufacturing will likely lead to another philosophical shift from even the JIT/TQC-based systems for which Japan is famous today. The need to create communication systems that keep the new kind of manufacturing enterprise integrated is a major challenge.[9]

Charles Savage has given considerable thought to the form of the new manufacturing. To describe it he has coined new terms taken from computer architecture. A *virtual memory* is one that provides the effect of a larger, physical, magnetic storage capacity by swapping blocks of information in and out of memory as needed. From that background, a "virtual team" is one that forms partly through assignments and partly through volunteers to take up assignments that the team itself may need to clarify in the beginning. "Virtual enterprises" are organizations that operate by composing and decomposing overlapping sets of virtual teams, some of them having members from other companies, to deal concurrently with multiple themes.[10]

Although a much longer evolution of these new organizational concepts is expected; teams, networks, and internal customers exist today. Companies are using them awkwardly, perhaps, but they are learning. One of the developments needed to assist progress is clarifying how to measure progress and performance in this new environment.

Figures 5–1 and 5–2 display simplified, conceptual views of the differences between functional organization (conventional) problem solving and a problem-solving *network.*

**FIGURE 5–1**
**The Hierarchal Problem-Solving Organization in Simplified Form**

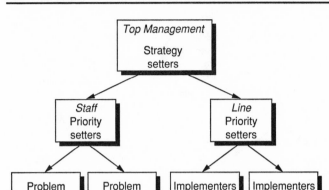

Figure 5–1 shows how staff problem solvers and line-working people are in separate silos connected at the top. Managers at middle levels set priorities and allocate resources to solve problems, often on the basis of overcoming the difficulties perceived in attaining their hierarchal performance numbers. That is, perception of problems is clouded by the method of organization, and by the functional separation of performance measures that go with it.

Most companies with major improvement projects form project teams or use matrix groupings to cut through the problems of functional organizations. Project organizations and task forces, while essential to making changes, do not overcome the basic biases of functional silos where those still exist as stout fortresses.

Figure 5–2 is intended to illustrate that teams of problem solvers attempt to cut through all the structure by including people associated all the way through an operations chain that stretches across many organizational boundaries.

Problem-solving teams may have both permanent and temporary members of their task forces, just as project teams assembled within a hierarchal organization. One difference is that improvement teams need more autonomy and more lati-

**FIGURE 5–2**
**Simplified Version of the Problem-Solving Network**

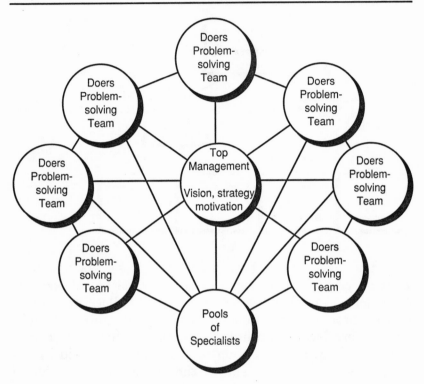

tude to react to customer needs than the "traditional" project team with objectives fixed by management. Since the teams are more free to establish their own objectives, they need a guidance system from top management to keep all the improvement effort headed in roughly the same direction. The top management guidance may in itself be provided by a top management *team*, and the teams need communication with each other to stay together, and to learn from each other.

The need to keep current in functional expertise does not go away, and not everyone will be on an improvement team at all times. The teams draw people from each other and from the specialist pools, which may be organized in a more traditional fashion; but specialists will hopefully not be kept in isolation for long periods of time.

Diagramming organizations for comparison while showing their multidivisional, multiplant, and cross-functional activities, is nearly impossible.

For example, in the problem-solving network, the decision makers and staff experts should be much closer to actual work than is usually the case in functional organization. The communication pattern in the problem-solving network should be broader—no strong chain-of-command procedures. And the "control process" for the problem-solving network is more difficult. In contrast, little could be simpler than a functional, top-down control system.

Some networks are merely informational: "Who do you know that has done a nice job reducing changeover times on heavy polymer mixing equipment?" "What is your team doing with 'X' company?" "Could we borrow some training material on...?"

A great deal of networking for various purposes goes on in functional organizations anyway, so what is the difference? It is that organizations structured to conduct regular operations are overlaid with various kinds of improvement teams. Most companies that are very far into continuous improvement have a somewhat degenerate functional organization for several reasons:

1. The existence of teams muddles clear functional distinctions.
2. Managements often abandon those measures that sustain a narrow functional focus.
3. People move across the functional boundaries regularly.

Will functional organization continue? Probably the vestiges will remain, but whenever possible, companies will organize small business units that focus on serving particular customers or market segments, or that concentrate on a given product line.

Most companies have not progressed very far at developing "permanently evolving" improvement teams. Most improvement teams are temporary.

Japanese companies are also making changes to further develop their own problem-solving, improvement networks. The typical large Japanese manufacturer (in Japan) differs from a counterpart Western organization on several points. The company itself is regarded as its people. Joining a company is a social bond more than a legal contract. Most employees join the company at a young age and stay with it. Once inculcated with the company's culture it is difficult to leave and be accepted in a new company culture. Consequently, over the years the cast of characters is stable, many employees know each other very well, and so the informal side of the company often prevails over the formal front maintained to the outside world.

The traditional Japanese company is a hierarchy by seniority. Seniority-first systems are slowly giving way to merit-promotion systems, however. Japanese line organizations are very formally structured. In some companies the supervisors even wear militaristic symbols of their station, but much actual decision making is through consensus and tangled webs of personal relationships.

In addition, the Japanese company also has a large number of improvement teams. Most have some form of quality circle activity in which most employees participate, and there are various cross-functional improvement groups as well. In some of the companies, improvement processes take place across tiers of suppliers through the various kinds of supplier-family organizations that the better manufacturing companies have developed.

However, a few Western companies are taking even more radical steps. So far as is known, Japanese companies are not experimenting with ideas such as self-directed work teams having no supervisors.

## POLICY DEPLOYMENT

In brief, policy deployment is a system whereby top management's vision of improvement is translated to more specific improvement targets all the way to detail levels in operations, and feedback from these levels helps to adjust overall policy before commitment to an overall plan. Figure 5–3 shows a simplified version of policy deployment, and Figure 5–4 is a summary of policy deployment as it has been practiced by an American company, Florida Power & Light Co.

Comparing the two figures, there seems to be not much parallel, although both describe a very similar process. Figure 5–3 is a sketchy view of how the process might work in a relatively simple, single-unit company. With few layers of operations, the top management of a small company can throw out proposed policies and targets for a coming year to everyone, and evaluate the responses from each improvement team first-hand. In fact in one company, Zytec, the president receives the proposed work plans of all the improvement teams in the company directly. There are no intermediaries.

In a company with thousands of employees scattered over a wide area, such as Florida Power & Light, the policy deployment process becomes more cumbersome, as shown in Figure 5–4. There are several stages and steps, but as with the simple version, at several points one party "throws a ball" to another. The development of improvement plans by this approach is sometimes referred to as a "catch-ball" system. The top management (central management may be a better description) tosses general policy proposals out for evaluation and planning. Various teams and departments toss their responses back. After a round or two of refinement, the company is off and running with a mass of detailed improvement plans by numerous people all working to support the overall policy. Hopefully the overall policy was established to make the company highly competitive.

While policy deployment is becoming well-known by that name in the United States, Professor Kano takes pains to point out that the Japanese term for it, *hosin kanri*, does not precisely translate to "policy deployment."[11] As practiced, policy deployment is really a way for top management to diagnose the ills of a total company and project the longer range, overall targets of improvement to which the entire organization should strive. It is a system of planning improvement that translates strategic changes into detailed activities. It

**FIGURE 5–3**
**Policy Deployment in Custom Fastener Co.**

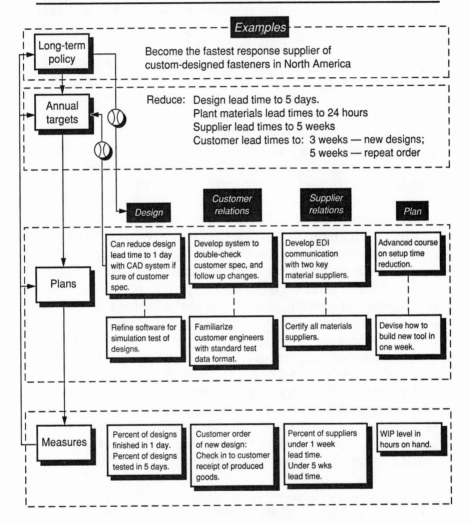

depends on people in many positions in a company having a good grasp of overall operational interrelationships within a company—understanding that, for instance, reducing tool-repair lead times is going to help cut material throughput times.

**FIGURE 5-4**
Overview Diagram of Policy Deployment at Florida Power & Light Co. as Presented in *Target*, Fall 1988, p. 8.

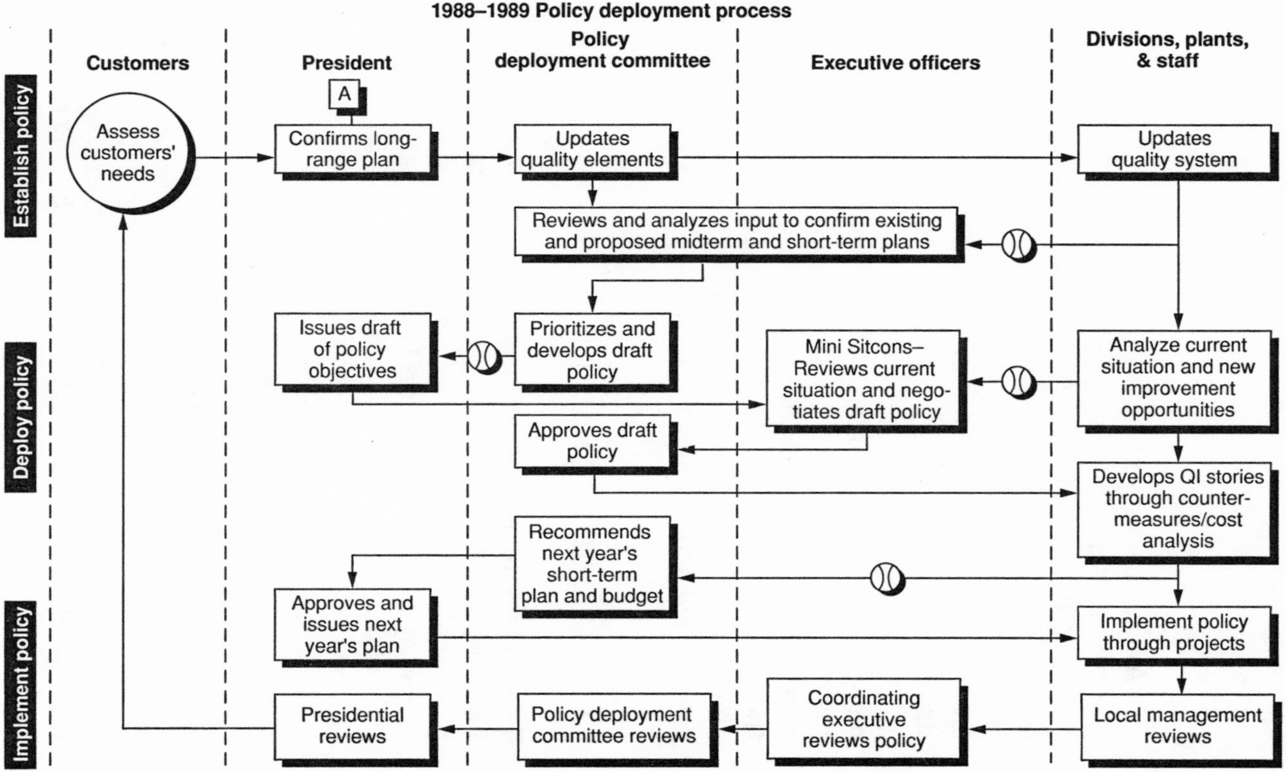

135

As can be seen from Figure 5–3, the kinds of measurements used with policy deployment may be cost figures, or if the objective is lead time reduction, the measures are time-based. Most important, *measures are derived from the nature of the improvements to be made.* The management (or others) does not permit objectives to be guided by a measurement itself.

In fact, the policy deployment process has an improvement process embedded in it. The feedback mechanism allows the objectives to be tested and refined by a process similar to that used for more detailed improvement processes. In fact, the Deming Circle (PDCA) logic was included in the process diagrammed originally in Figure 5–4, but was removed because it created too much clutter.

## LEADERSHIP IN TRANSITION

The leadership necessary to guide an organization through an improvement process is easily underestimated. Once a leader thoroughly understands the logic of the new manufacturing, it is easy to believe that surely everyone else in an organization will catch on quickly, and enthusiasm for the new way will be easily transmitted. Actually, leadership takes an enormous amount of people contact, recognition, example-setting and direction-pointing—leadership energy.

Directing a manufacturing enterprise merely engaged in routine operations is comparatively easy. Sometimes it can be done almost exclusively through a measurement and reward system. Leadership through budgets and top-down controls requires little persuasion and interaction, so leaders may vary considerably in personal style. Being a champion of a continuous improvement process requires advocacy.

The usual expectation of incentive systems is that by tying monetary rewards and other recognition to the personal achievement of each individual, all employees will be motivated (sort of a personnel counterpart to the concept of separately measuring the cost performance of independent operations). If these measures narrowly focus an individual on maximizing output, they are clearly unbalanced, and even those measures that concentrate attention on the performance of a person's functional department contribute to the fortification of functional silos.

To overcome all this, personal performance measures need to be connected to overall improvement, and individuals should be encouraged to cooperate as team members. One approach is to establish the measurement process itself so that the customers of a team have a big role rating the improvement performance of the team, and within the team, peer rating indicates whether individuals are cooperative and contributive to the team.

Without doubt, life is much different when performance ratings are dominated by internal and external customers and by peers, rather than bosses. The reward systems, status systems, and the measures themselves, can all be expected to differ. American companies are just beginning to venture into this new territory.

Companies deeply committed to employee involvement are also deeply committed to systems of widespread recognition that stimulate united improvement effort. Coupled with the catch-ball system of improvement guidance derived from policy deployment, the necessary style of leadership is quite different from the top-down, functionally divided approach to management, and the performance measurement system is also different.

One objection to the "empowerment" of decentralized improvement processes is that employee recognition is used manipulatively. This contention assumes that the real motivation is profit to the ownership and bonuses to the managers, making the recognition hypocritical. In some cases that is undoubtedly true, but not in all.

However, *every* system of personal performance measurement is manipulative in some way. Any measurement intended to influence behavior has some degree of manipulative intent. Acceptance depends on whether one agrees with the direction of the improvement process. Employees taking initiative pleasing their customers is less manipulative than top-down, departmentally focused, carrot-and-stick systems.

Measurement systems cannot substitute for leadership in the new manufacturing. Trying to find a set of performance measures that will make a philosophical and motivational transformation without leadership effort is a search for the Holy Grail. The measurements and measurement systems are not constant. They change and evolve as a company changes and evolves on its way to something better.

## NOTES

1. XEL Communications, Inc., Aurora, CO, is a small custom manufacturer of printed circuit boards.
2. This phrase originated with Phil N. Ensor, "Organizational Renewal—Tearing Down the Functional Silos," AME Study Group on Functional Organization, *Target* 4, 2 (Summer 1988), pp. 4–14. Few meetings or workshops on JIT/TQC today end without some discussion on new ways to organize.
3. José Ortega y Gasset, "The Barbarism of 'Specialization'," as collected in his *The Revolt of the Masses* (New York: W. W. Norton, 1932). (Translation from the Spanish, originally published in 1930.) This essay is a favorite of liberal arts scholars seeking ammunition to deflate egos in the cross-disciplines.
4. The controller of an Indiana company, in an unguarded moment, captured the essence of the conflict with the remark, "Shrink the inventory? God, we need it to adjust profit to where it needs to be!" In another incident, a production manager, complaining about the vagaries of short-term efficiency measures, left the meeting to switch his personal funds from one money market account to a better performer.
5. Barry Schwartz cites several pieces of evidence in a book called *The Battle for Human Nature* (New York: W. W. Norton, 1986). He notes that a school of economists typified by E. J. Hobsbawm has mentioned that in the pre-industrial age, economics was intertwined with the social relationships called feudalism, by which neither land nor labor could be sold, so the possibilities for money-based reward structures were extremely limited. The work of serfs was menial, but usually varied, and methods generally were hand-me-down rather than scientific.

   It took a long time for the industrial revolution to break down this old order. Many early factories had difficulty retaining workers to do repetitive work, and until the age of scientific management, there was no custom of rewarding workers with higher pay to produce more output by a prescribed method. Thus began short-term performance measurement for workers—and for their bosses.

Schwartz cites the economist Stephen Marglin as attempting to logically demonstrate that division of labor was not done to promote productivity. Division of labor and compartmentalization allows the bosses much greater *control* of the entire process. Marglin knew nothing about eliminating waste by fast-flow production, but did realize that substantial human energy spent in measurement and control could be more directly engaged in value-adding activity. Much of the control is exercised through short-term, narrowly focused reward systems. [Marglin's reasoning is in "What Do Bosses Do?" in *The Division of Labour*, ed. A. Gorz (London: Harvester Press, 1976), pp. 13–54.]

Finally, Schwartz recounts a stream of research on behavioral reinforcement as demonstrating that reinforcement has two effects: First, it gains control of an activity. Behavior that people might have enjoyed anyway they will pursue with even more enthusiasm if they are rewarded for it. Second, if the reinforcement is withdrawn, people do not return to their natural enthusiasm for games or puzzles. Activity that once seemed to be fun is turned into drudgery. (Several such experiments are found in *The Hidden Costs of Reward*, ed. M. R. Lepper and D. Greene (Hillsdale, N.J.: Erlebaum, 1978).

One of the most interesting experiments is Schwartz's own. College students are asked to play a game in which they push buttons to illuminate lights on a board. The direction to go is determined by which button is pushed—a form of video game. Going in a correct direction lights a light. An unknown pattern of lights exists.

One group of players with no prior experience playing this game was asked to discover the rules of the game, that is, all the sequences that contribute to discovery of a total pattern of lights. Some were paid by the light. Some were paid for each rule discovered. Some received no payment at all.

For this group of players the reward system had no observable effect on behavior. All players varied their play in an experimental fashion to discover the various patterns of light. Almost every one discovered all the rules and did so rapidly. Reinforcement had no effect.

A second group of players was preconditioned by playing the game one thousand times under the rule that every light found is rewarded by a small sum. During the preconditioning, all players found one routine and went through by rote, often while thinking of something else. When this group was then asked to discover the rules of the game, they acted disinterested, as if discovering the rules mattered little to them. They were significantly slower and less efficient in the discovery process. Schwartz concluded that the single-focus, immediate-reward pretraining dulled the students' normal enthusiasm for problem solving.

Whether laboratory experiments on students and other convenient groups of people have any validity when applied to manufacturing companies is, of course, always open to questions. The hypothesis to be derived from the laboratories is that short-term, repetitive reward systems have a lasting effect that inhibits people subjected to them from vigorously contributing to continuous improvement. It would be useful to have field research to substantiate the unsystematic observations of many people that measurements such as many of those in Figure 5–2 are stultifying.

6. One of the few interesting results of the author's dissertation was that from a sample of 138 scientists and engineers selected across the full spectrum of supervisory performance ratings, only three could self-rate themselves below average. This finding was not unique.

7. Every worker being a manager is not a new idea. Scott Myers wrote a book by the title *Every Employee a Manager*, 2nd ed. (New York: McGraw-Hill, 1981).

8. Tom Peters, *Thriving on Chaos* (New York: Alfred A. Knopf, 1987). Robert H. Hayes, Steven C. Wheelwright, and Kim B. Clark, *Dynamic Manufacturing* (New York: Free Press, 1988). Rosabeth Moss Kanter, *The Change Masters: Innovation and Entrepreneurship in the American Corporation* (New York: Simon & Schuster, 1983).

9. The Japan Machinery Federation, *Survey Research Report on the Development of New Manufacturing Systems*, March, 1988.

10. Charles Savage, *5th Generation Management* (Maynard, Mass.: Digital Press, 1989).

11. Noriaki Kano, "TQC in Japan and the United States," GOAL Sixth Annual Conference, December 5, 1989, p. 8.

# CHAPTER 6

---

# MEASURING UP

---

### On Leadership of a Change
### to Operational Excellence

"The journey involves the difficult task of reshaping people's thinking so that they approach their jobs with this process ingrained in their minds. This is what it takes to 'get from here to there.' It is a difficult journey that really never ends. And outside of internal recognition, there are no accolades along the way."—Marshall McDonald, President, FPL Group, Inc. (Florida Power & Light) Excerpted from 1987 Annual Report.

When leaders are trying to turn the direction of a company, their effort is typically accompanied by a shift to unfamiliar performance measures. Someone uncommitted to quality will protest that the company is measuring quality just to be measuring quality and neglecting other matters. Why not depend on the old, more familiar measures "so we can know where we really are." There is great danger in these arguments. They represent the first steps down the slippery slope into a hodgepodge of directionless measures that undermine leadership for improvement.

Using measurements in the improvement process is the vital concern, and they should complement leadership. Measures in the absence of leadership are sterile. Leadership in the absence of measures is robbed of a primary medium of communication.

No single set of performance measures fits every company at every point in its manufacturing progress. Far too many issues and problems permeate measurement for that. By way of illustration, most manufacturing consulting firms have proprietary checklists for evaluating prospective client companies. These lists are not all the same. They vary depending upon what aspects of performance the consultants themselves consider strategically vital, the services each firm can offer, and the degree of investigation possible. Some of the lists are simply ratings based on quick visual impressions; others ask for hard, credible data.

A process to select a best-performing plant or company brings out many of the issues. It is not easy to compare different companies in different industries; excuses about being different have no end. Selection is only possible if judges are firmly agreed on the philosophy of improvement the contenders should be pursuing, and on the performance criteria by which comparisons can be made.

Directional agreement—compatibility of goals—leads to consensus on criteria, and the criteria suggest metrics. For example, if customer satisfaction is a goal, customer complaints would be a criterion, and the number of customer complaints, sorted into a Pareto classification by causes, would be a metric. However, in the close calls, metrics only help sharpen judgment. A score tallied from a long list of measurements does not constitute a conclusion about which of two plants might be better, regardless of the purpose of the comparison—friendly competition for mutual improvement, make-buy decision, or other reasons.

Performance measures should not substitute for thought. Given a hint that decisions are made only by the measurement results, people may work to post numbers rather than actually strive for the goals they represent. For example, someone measured by customer complaints may actually cajole customers into accepting poor products and service rather than improve them, particularly if that single number seems to dominate managerial judgment. Just as obsession with efficiency numbers can cause production managers' imaginations to go numb, turning other metrics into icons can create the same dysfunction in others.

The challenge is the use of measurements in the improvement process. No one questions the value of many traditional business measures such as cash flow, but the most exciting measures are those describing whether customer satisfaction and operating capabilities are improving. While the companies cited earlier—Harley-Davidson, General Electric, Xerox, and others—can doubtless do *more* to align performance measures with goals, they all share the intent of doing this. No performance measure should distract anyone in the company from the new manufacturing.

## THE MEASUREMENT PROCESS

Just as one set of performance measures cannot cover every situation, one book cannot cover all the conditions and possibilities of a measurement process. The performance measures are too deeply embedded in organizational work at every level. We can only present an overview of some measurement fundamentals useful for improvement of manufacturing performance by the new paradigm.

As an *integral part* of operational improvement the measurement process consists of first defining measures relevant to desired improvement, then using them to gather, analyze, and take action based on the data—management by fact. And for this purpose, every worker is a manager.

### Source Data

Tracing to the original source of data penetrates through even the most complex measurement systems. From what source data are other measures derived? When, where, and how is basic source data obtained? Examples of basic source data might be: counts of parts, weight of resin in a drum (or more exactly, the full and tare weights as captured by a weighing process), elapsed time taken between two readings on a clock. Floor dimensions taken with a tape measure.

Many performance indicators are ratios, valuations, classifications, or combinations of basic source data, and all should start from reality. Suppose we chart the flow of a part through a plant, from truck entering the gate to truck leaving the gate. The first abstraction is marking a chart to denote any apparently different state of the part. A second level of abstraction is coding different states into classifications: moving, being transformed by an operation, sitting in stock, and the like; then summarizing the totals of these categories. If time in each state is observed, more source information is recorded. If we identify the value-adding steps in this flow, another abstraction occurs. Calculating a ratio of value-adding steps to the total is another abstraction. A time ratio, that of value-added time to total time in the plant, is another abstraction, and so on.

Most performance measurements are built up from source data. Familiarity with the measurements dulls the consciousness that the process of obtaining the source data may not produce measures that represent reality as closely as assumed. And yet our reasoning process works because we are able to abstract from reality, and the measurements are a natural part of thinking.

Appreciation for the remoteness of many performance measures develops respect for the viewpoint that it is better to simplify a process by direct observation before analyzing it by abstract measures laboriously compiled. Simply do the obvious first. That is the reasoning behind the now well known practice of Workplace Simplification, also known as "5S": (1) remove all excess items, (2) standardize locations, (3) clean, (4) practice discipline, and (5) extend participation to everyone.

Despite the risks from distortion of reality, abstract analysis has the potential to create insights not apparent by direct observation. One option, used when improvement ideas seem to be lagging, is to start over, rethinking sources of data and beginning a fresh analytical build-up. But this is not done very often because of human "inertia." It consumes too much time and energy.

## Basic Metrics

Many managers learn a few basics of metrics even if never formally schooled in them. The most basic ideas are the concepts of precision, accuracy, and overall error, but include the issues of defining a measurement itself. The concepts mix together. Basic metrics should be taught in high school, but frequently are not, and even people who understand them well can forget about them in the press of responsibility.

## Defining a Measure

The relationship of a measure to a performance criterion is not always obvious or easy to establish. Measurement definition must both consider the problems of obtaining data, and the necessary precision and accuracy of the results.

As a physical analogy, consider measuring the sizes of an assortment of particles. First comes the problem of the range of sizes and the degree to which the sizes must be categorized. If the particles range up to gravel size and are hard, they can be sifted through a series of screens. If they are micron-sized, the same idea may work, but requires a different technique of forcing material through membrane-type filters. Or one might become involved in the problems of visual comparison with a standard in the measurement process.

Defining the measurement approach is not independent of clarifying our purpose for making the measurement. Do we need to know maximum or minimum dimensions of particulates and thus begin to describe shape? Do we want a measurement indicative of total volume, and therefore mass of individual particles? Will a rough average do, or do we need to capture a distribution of sizes?

"Management" measures must also be carefully defined. For instance, the term "lead time" is used casually, and is often called "cycle time." There are lead times for many kinds of activities, but one of them is throughput time, the time required for material to flow through a production process. That definition must be further clarified for measurement.

Is it gate-to-gate time, record-of-receipt to record-of-shipment time, or first-process-start to last-process-finish time? Differences can be substantial.

Then do we want a rough average, or do we need to know the throughput time of specific parts, or specific classes of parts? Watching and timing a specific part, step-by-step is diligent work—more work than most are willing to do unless convinced that such a measurement is necessary. If a specific unit can be marked all the way through the process, and if all that is wanted is overall elapsed time, the time from a beginning point to an end point can be clocked. Many parts must be timed if a distribution of times is desired, and that becomes feasible only with an automatic check system that uses markings such as bar codes. All these methods take resources.

Fortunately, there is seldom a need to have more than a rough estimate of throughput time. In that case, take a count of all the parts between the two points defined, and divide that count by the average rate of use of the same part. If a more precise, detailed measurement is taken, there should be a reason for its necessity. Perhaps we want to see whether some parts are getting "stuck" in the process—never moving on through—but it is probably easier to check for this in another way. Sometimes it is desired to have a low variance in the throughput time of items that undergo persistent chemical change, for instance. Even then it may be easier to simply check whether discipline to maintain work in sequence exists for the work-in-process.

Measurement definition is the first basic. Measurement is defined by the practicalities of measurement and by the degree of need for precision, accuracy, and clarity of the results.

## Precision and Accuracy

Precision is the repeatability of a measuring process, demonstrated when multiple measurements taken the same way are nearly the same, or in statistical terminology, if differences between measurements exhibit a small variance. Accuracy is a more difficult concept that roughly implies that the result of a measurement process is very close to a true value, if a true value can be established. Although perhaps tedious, understanding precision and accuracy is important in the measurement process.

To be defined a little more scientifically, accuracy is the probable error between a measurement and a standard. A standard is a precise, accepted measurement of the same thing or a similar thing. In the case of physical measures, standards are established by such institutions as the National Institute of Standards and Technology (NIST). On the technical side of quality control, it is important that companies maintain all instruments and be able to trace the calibration accuracy of all critical instruments to those of NIST, or similar bodies. Arguments about the calibration of two instruments used to measure the same thing, but reading differently, are too late when disposition of expensive material is in question.

Accuracy measurements should take into consideration both bias and precision. Bias is the difference between a measurement-derived estimator, usually a mean measurement, and a "true value." For those technically serious about accuracy, a description of accuracy should include the source and method of measurement, the standard used, and some quantification of both bias and precision; and even then accuracy may not be a given because different analysts may interpret the same data differently.

The term "accuracy" is often used loosely when applied to management's measures, often meaning only that a measurement method, or the result of it, agrees with the speaker's reasoning. Even when interpreted with a little more rigor, "accuracy" seldom means that we can make a probabilistic statement asserting how close a measurement is to some true value.

For example, "inventory accuracy" generally means that an inventory record agrees with a count. Therefore the number is assumed to be correct. What is really meant is that successive counts of inventory have been precise. There is no way to know if the same systematic error was made on the last count as might have been made on the count before—and used to adjust the record. A systematic bias in counting is easy to have in the event of overlooking hidden items, of assuming boxes are full when they are not, or in other ways. Whether a count was technically accurate may never be known.

For practical purposes, precision is a good indicator of accuracy in many source measurements, such as counts, but a rigorous concept of accuracy is also good to keep in mind. It prevents acceptance of silliness.

Assuming that a long string of digits implies either accuracy or precision is a common error, even among people of experience. Computers and calculators enable us to easily carry computations to 8 or 10 significant digits. For example, say a sack of apples I pick from a tree weighs "about" ten pounds, just by guess. I scoop up about two dozen apples, but eat one. By the calculator, the weight per apple is 10 ÷ 23, or 0.4347826 lbs. per apple. Computers impart an aura of precision or accuracy by spilling out digits from calculations using data taken from the crudest of source measurements.

There is little time to question every number presented, but workers (and especially managers) need to question measurement processes and beware of the deception of elaborate computation and presentation.

Suppose that a dielectric film comes in sheets, but is sold by the pound. An analyst weighs some, and checks the thickness of several sheets. Using a calculator, the analyst comes out with a thickness of 0.0246853 inches, and the bill of material cost will be \$2.7852846/ft$^2$. A moment's thought reveals how imprecise the source measurements must have been from which these numbers were derived, and the numbers are averages that say nothing about thickness variance, which is important in the application, or about any assumptions of process losses.

Precision and accuracy of variables are sometimes easier to understand than precision and accuracy of attributes. A dimensional measurement can be expressed as a *variable*, a number that can take on any of an infinite number of values. An *attribute* is captured only as an attributed number, such as a subjective rating for the overall appearance of an automobile. It could be scored using numbers from 1 to 5, a semantic differential scale ranging from awful to superb. However, if the same group of people rates two cars, and the average scores are 3.7 and 3.8, it is hard to assert that one is really better than the other, even if statistical analysis shows the difference in the groups' subjective measurements to be significant.

## Cost Inaccuracy

Most performance measures are built up by combining source data, and all costs are developed from source data using a cost model. As discussed in Chapter 4, there are many alternative cost models. *"Accuracy"* of a cost figure depends primarily on whether the cost model adequately incorporates source data from activities believed relevant to the purpose for using the cost data, and secondarily, on the precision and accuracy of the source data.

Discussions of cost accuracy become entangled with problems of cost relevance. *Relevance* of any metric is its closeness of association with performance goals and criteria of interest. A relevant cost should summarize resource use associated with a particular situation or decision, but relevance judgments depend upon knowing the situation. For instance, the relevance of a warranty return rate (and cost) to a product's early-life quality depends on the aggressiveness of product claims and their promotion, as well as on the actual early-life dependability and functionality of the product itself.

We can evaluate the precision and accuracy of some source measurements with a rigor that approaches scientific, but establishing a cost measurement as accurate in a scientific sense is nearly impossible. As generally used, cost accuracy implies one or more of at least three meanings.

1. Cost measures are precise. Different cost calculations for the same thing are in close agreement. Since precision is often a good indicator of accuracy, this is the most scientific of the three meanings.

2. A cost figure is calculated using an accepted cost model. Usually this means that a cost figure was developed by the organization's "official" cost system, so the number thus determined is held up as standard even though a "true" number cannot be scientifically estimated, and arguing with the logic of the system would be wearisome.

   More broadly, a cost number is accepted by *model confirmation*—the cost methodology appears "reasonable," and therefore supports judgment that the resulting number is "accurate." As an organization begins to make decisions using cost models tailored to separate purposes, the practice of cost acceptance through consensus on model relevance is an issue that takes on more importance.

3. Cost figures agree with judgment, even if the cost model is not explicitly understood. This is sometimes called *results confirmation*. If the results of a cost computation agree with the intuition of those familiar with the process, much less explaining is necessary, and arguments should be few.

Unfortunately, as activity-based cost methods and other costing alternatives begin to be used, cost accuracy is indeed most often established by checking whether results agree with intuition. The good news is that early experimenting with cost models shows that if model designers share a common philosophy of relevance when designing their models, results do not widely diverge. If the majority of cost users can accept cost results, the system is accepted.

The value of a cost number depends on the cost model used: definition of relevant activities, cost drivers, cost objects, and methods of allocation. Numbers from different models can be substantially different, as was seen in Chapter 4. Cost model users should be aware of the distortions from the models used. For instance, a unit cost model that simply allocates engineering cost on the basis of direct labor grossly distorts the cost of a product built mostly by automation, but requiring the lion's share of engineering effort. For easy understanding and discussion, most cost methodologies should be simple, but if they are simple, they are also apt to have some kinds of inherent biases that should be considered when using the data.

One of the most common deceptions is mistakenly associating the logic of the cost model for the logic of the process—or for the guide to short-term improvement of a performance number. Anytime an overhead pool or fixed cost is divided by unit volume to get a unit cost, it is easy to be tempted to depress the unit cost by increasing the volume number in the denominator—whether output is needed or not; so a new, expensive autoinserter *begs* to be loaded with work to spread the capital expense, for instance. Another common deception is believing that a cost driver, such as labor in the "traditional case," is somehow much more important than it really is because its valuation has been inflated by using it to allocate *unrelated* costs.

Many costs are also far removed in time and abstraction from the source measures from which they were compiled. Consider the development of a unit cost to heat-treat a part in a furnace. Most people would agree that there are at least three resources to be considered: (1) cost of furnace, (2) upkeep, and (3) energy. Cost of movement, control, and so on, will not be included to keep it simple. The basic, source measurements in each of these categories are subjected to questions of relevance, precision, and accuracy. Then they are apportioned to parts by some logical approach. If a decision hinges on a difference between costs thus derived, it takes a huge contrast to provide conclusive evidence.

Suppose the decision is make or buy—to outsource the heat treatment of one part. A prospective subcontractor submits a bid stating a cost per part, and it is a "hard" cost, payable by check. The internal cost is partly developed by "imputation," so it is "soft," but bigger. The temptation is to accept the hard cost and to outsource, but a series of such decisions takes one into that old trap—increasing unit costs by decreasing unit volumes. As volume for the in-house furnace drops, the apparent cost per unit rises, thus leading to more outsourcing, and so on; a "death spiral" eventually leading to abandoning in-house heat treating. Meanwhile, as long as the furnace is used every day, resources consumed by it are little changed.

The death spiral in unit costing is really another version of allowing nonrelevant cost models to dominate a decision that should be more strategic. That is, the decision should review part designs, skill maintenance, lead times, quality, and other factors. Possibly the biggest problem with cost models is simply accepting measures used in decision making with little question of how they have been developed or what their intrinsic biases in structure are. People who would pin the accuracy of a physical measurement down to the nubs will make decisions using cost figures of dubious origin without incorporating the uncertainty into their judgment.

Hopefully these tendencies can be curbed in the future by cost users having more ability to model costs from source data themselves, using different models for relevance to different situations. For instance, decisions on short life-cycle projects should use a life-cycle cost model. To use costs well, users should discuss the biases of cost models more openly, and better understand their effect on decisions. One can even expect discussions to incorporate costs calculated by several different methods—along with caveats about each model from which cost numbers came.

## Timing of Measurements

The sins of quarterly financial reporting have been soundly denounced by almost everyone interested in improving manufacturing. The sin is not really the quarterly measurement itself. It is *impatience.*

A proposal to legislate a ban on reporting quarterly financial results has been floated. The absence of such reports will remove one mechanism by which trader impatience is transmitted to corporate operations, but it will not do much for the impatience itself.

Deciding how often to take measurements is another matter. Any answer is based partly on the practicalities of measurement, partly on how often anyone can take any action based on a measurement, and on how often measurements are useful to detect variations in a process. There is no general answer to the timing question, as with many other questions. A thermal controller on a machine is monitored constantly. Growth of a hardwood forest needs little checking—but frequently enough to detect disease in time.

Most people are interested in the timing of major management measures. Timing should depend on the "natural" work cycles of a business. The harvest cycle is an obvious planning and measurement period for a farm. Inside a plant some long unit work cycles are easy to spot. If a plant builds one locomotive a week, it has a natural one-week cycle to its work. If it builds several locomotives per week, the time between replanning is more uncertain. Many companies drift into a month as a "natural" improvement cycle. That is, once a month there will be a major change of schedule and perhaps of work content. With the shop revision occurring at that time, new improvement ideas may be incorporated.

Sales work has cycles too. How often should one call on a customer? How often to follow-up to see how products are working? There are no pat answers, but the measurement cycles need to provide good data for effective redirection of effort at the right times. It is also very helpful if production-change cycles, sales-planning cycles, and accounting cycles are on the same wavelength, so to speak, but very few companies give thought to such matters.

Measurement frequency and measurement summary cycles depend on the timing of the changes and corrective action that should be taken in response. Measurement is communication, so measurement cycles depend on planning and execution cycles. Measurement summary cycles fall into a continuum of categories:

| *Timing* | *Plan or Decision (Example)* |
| --- | --- |
| Irregular | Plant, company, or market territory expansion. |
| Product life cycles | Product design, process design, and product launch. |

| | |
|---|---|
| Schedule changes | Weekly and monthly changes in production volume and mix. Also changes in sales-plan emphasis. In an excellent company, schedule changes are good times to further improvements—incrementing with each schedule change. |
| Quality cycles | Times when machines should be adjusted, tools changed—or customers be checked. |
| Constant monitoring | Fail-safe checking, often automatic. |

## Making Comparisons

There are several desirable features of measurements. One is that they provide guidance—even inspiration—but to do this, measurements must be useful making comparisons. Another is simplicity. Simple, easy-to-understand measures also promote discussions and comparisons. In addition, simple measurements do not consume many resources in the data-taking itself.

Measures considered should be those relevant to a decision. Relevance is most often considered as a property of cost measurements because they combine other measurements. Relevance opens up *all* the issues of cost models and systems, but it also begs the analyst to consider all the issues (and measurements) relevant to a decision, not just cost numbers.

Recall the kind of decision represented by outsourcing a part for heat treatment. The issues are quality, lead time, flexibility, capacity, dependability, worker relations, and supplier relations—as well as cost. Judgment in the absence of data may simply be that the supplier is comparable in all the performance criteria, and there may be precious little time or reason to analyze deeply. If parts are needed *now*, and other factors do not veto outsourcing, a decision is easy, and probably temporary.

A more permanent decision should consider a review of *all* relevant factors. It's generally better, if comparative measures are considered necessary to the judgment, to use separate measures for each factor. There is a tendency to try to make one measurement do too much. The desire to express everything in dollar terms is something of an economist's penchant, as if converting all other reasoning and experience to a price-cost-quality basis will somehow clarify matters not otherwise very well understood. Typical examples are: cost of holding inventory, cost of quality, and opportunity costs. One can imagine valuations of dependability, opportunity costs of lead time, costs of skill deterioration, ad infinitum. One that is out of favor is valuation of pleased customers, called goodwill when capitalized on a balance sheet. It is rarely allowed by the Financial Accounting Standards Board. Pleased customers are also possible to *displease*.

Dollarizing all measurements is sensible only if one believes that every action taken can be related to profitability. The downfall is ignoring the factors of the business that contribute to technical strategy or customer satisfaction—or dealing with them in such an abstract way as to be ineffective.

It is not necessary to convert everything to a dollar measurement in order to make comparisons. Consistency of definition and of measurement method are necessary to make comparisons. There are three basic kinds of comparisons: (1) with the same unit's past performance, (2) with another unit's performance, and (3) with some standard of performance otherwise obtained.

Consistency raises issues of the management of measurements. Should a company require identically defined measurements from each plant? Probably it should not require *completely* identical ones. The same measures do not mean much when compared between an assembly plant, a fabrication plant, or a chemical plant. Even when comparing two electric motor assembly plants, every measure may not be identical. One may assemble small motors for large-volume customers; another large motors with options for small-volume customers.

However, after all the cautions, comparison is the essence of using measurements. Progress is only verified by comparing the same measurements over time. Much can be learned by comparing similar operations in two different locations. Comparison for improvement's sake is important.

## DRIVING STAKES INTO THE GROUND: MEASURING IMPROVEMENT

Our primary interest is measuring improvement; stimulating it rather than stifling it. This purpose intersects all other purposes of measurement, but a comprehensive review is impossible. Voluminous measurements are used for many different purposes: investment analysis, environmental protection, and fair employment practices are three among many. Measures for improvement should track progress toward desired goals. Some of those may involve investment, environment, or employment.

Sorted out differently, major improvement goals are in six categories: (1) quality, (2) dependability, (3) waste (resource) saving, (4) flexibility, (5) innovation, and (6) development of people. An overall set of performance measures should relate to all these goals, although no company can summon the energy for intensive improvement concentrating on all areas at once.

### Trend and Context

Measuring trends in actual performance is very important. This is really what is meant by the phrase popular among production managers leading change, "driving a stake into the ground" to see if we've moved. When deeply involved in the intricacies of continuous improvement, it is hard to know how far one has come and how fast one is going. Posting some primary indicators for everyone in a company to see is a good idea.

Communicating a performance measure is only made harder if it is complicated. Even if a performance measure is not defined to be closely aligned with a goal, charting a simple one allows most people to easily relate to progress. Few people need anything more than Figure 6–1 to interpret the progress shown. The graph shows the same measure over time. There is no indexing, no change of scale, and no question which direction is better.

Figure 6–1 also displays context. The trend line shows how recent performance compares with the past, and the prominent target markers also show how it compares with goals and with a level of performance discovered elsewhere. At a glance, everyone gains an appreciation where performance stands, and where it is headed.

Defect rate variances from the targets are displayed in Figure 6–2, but suppose a cost variance had been shown. The purpose of this chart is to illustrate how difficult it is to comprehend "where one is" by posting differences from a goal, or variances, as they are called in costing. Difference measurements are one more complication cluttering interpretation.

The target values displayed in Figure 6–1 may be somewhat arbitrary, but even so, they help establish where we are in comparison with where we would like to be. Targets and benchmarks on a trend chart help establish context in the

**FIGURE 6–1**
**Displaying an Improvement Indicator**

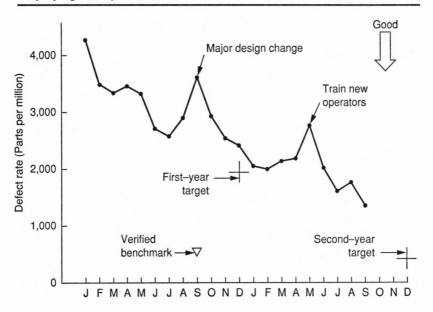

**FIGURE 6–2**
**How Not to Display an Improvement Indicator**

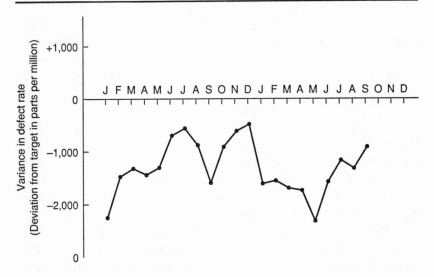

same way that finding a landmark on a map helps a lost traveller to visualize the location. Figure 6–3 is an example chart from Florida Power & Light Co. It is interpretable by most people without explanation.

Context also means relating the measure to the process being measured. Just stating the measurement as a defect rate without describing the process allows no one to compare the measurement with their own experience—if they have similar experience to relate to. Suppose we describe the measure in Figure 6–1 as the defect rate just after wave soldering for a multilayer printed circuit board using through-hole component insertion. Anyone having the slightest association with a similar process begins to make associations and comparisons. Given the knowledge that the same plant expects to start switching to surface-mount technology at the start of the third year also helps put the situation in context. This same group of people will soon be starting a new trend chart following a major technical change.

The learning rate on surface-mount defects should also begin to show a downward trend. From the viewpoint of im-

**FIGURE 6–3**
**Distribution Service Unavailability (An Example of an Actual Performance Indicator Courtesy of Florida Power & Light Co.)**

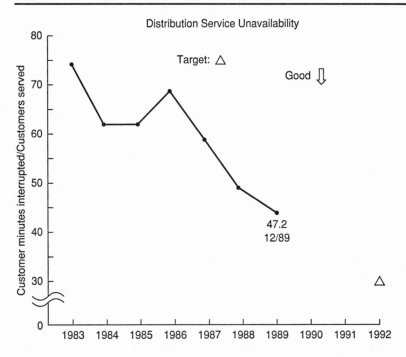

provement processes, the capability of people to make improvement is really the constant that is expected. No one has an unbroken improvement trend forever. Operators, machines, designs, and technology change, thus interrupting the improvement process. But if continuous improvement has become deeply ingrained in an organization, soon after a change, the trends in appropriate improvement indicators should resume a favorable direction.

## Linkages

Performance improvements are interlinked. A company cannot long continue to decrease lead times without beginning to seriously address quality issues in an organized way. A company that begins to greatly improve quality soon finds that resolving quality issues reduces lead time, and that from the viewpoint of the next operation as customer, short lead times equate to better customer service.

The same interconnective logic carries through a great deal of manufacturing excellence and the performance measures typical of it. Take the many reasons for reducing setup time, for instance. The most obvious is to reduce lot sizes, and through that, to reduce lead time of flow through a production process. However, smaller lot sizes also mean more setups, which means that the workers must avoid waste during setup by doing it right the first time. Then the quality of parts from the setup should not be in question.

There's more. For quick setup, both the equipment and the tooling should be well maintained. If tooling is prepped by a tooling shop, the lead times in the tooling shop should also be short. Thus begins the same round of improvement linkages in the tool shop as on the main floor. Likewise, equipment maintenance becomes more noticeable. If not already started, a program of preventive and predictive maintenance must begin.

To improve setups, they must be practiced regularly. Turning the setup process into a routine is the essence of improving it, but in order to practice setups regularly, production schedules need to permit this regularity. That leads to uniform load schedules with *pull system* connections drawing material in the repetitive-case companies.

In short, the logic of production system interactions and supplier-customer linkages leads to performance measurements combining and pointing organizational improvement in a strategic direction, although the leadership must maintain the direction.

### Key Performance Indicators

Performance measurements are not all equal. Some are key indicators that upper management should use to see if the company is moving in the desired strategic direction. A balanced set of key indicators should cover all six improvement goal categories. In some cases, an indicator covers more than one goal.

Figure 6–4 is an example of some key indicators used to compare three divisions of the same company selling in different market segments. A glance is sufficient to confirm that the performance ranking of the three divisions: A, then B, then C.

A key indicator cannot be improved without actual improvement taking place in a wide spectrum of company activity. Key indicators should represent progress on a major, company improvement goal. In Figure 6–4, "Quality ranking" and "Fallout" rather obviously fulfill that function of measurement. The others may not be obvious.

"Total productivity" is a measure of the value added per person. Many companies have had trouble assessing performance with this measure. Value-added productivity can improve or deteriorate in several ways. Adding new products with higher margins will increase it. Outsourcing production work will decrease it unless the advantage of outside production is truly substantial. Decreasing the number of people needed for the same activity will increase it.

A top management that deliberately makes key use of a value-added productivity ratio should think through the implications of using it, and be sure that this or other measurements, reflect exactly the policies they believe will make their company competitive. It is a mistake to promote measurements without promoting the policies.

"New model development time" in Figure 6–4 is an indicator of a company policy to compete with new products. It is an indicator representative both of flexibility goals and innovation goals. It does not state whether the company intends to be a technical leader or a fast-follower. However, reducing model development time is difficult to accomplish unless there are improvement efforts to reduce lead times in many other areas. Among them:

- Throughput times, setup times, and paperwork times in production. A new product cannot be quickly introduced into a mediocre production organization.
- Engineering-change lead times must also be short.

# FIGURE 6–4
## Key Performance Indicators: Three Electronic Instrument Divisions

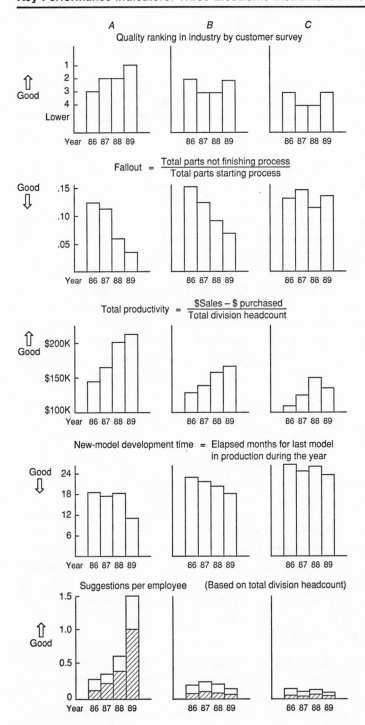

162

- Preproduction lead times must be short. Short product lead times are impossible without short lead times in tooling and equipment preparation shops, for example.
- In general, both people and operating methods must be flexible. For instance, time to change a layout should be short. Time to increase production volume should be short. Time to change production mix should be short. The company should have successfully decreased many lead times-to-change.
- Design-change lead times from suppliers must be short. This implies an ongoing program of quality, lead time, and technical improvement in cooperation with selected suppliers, and that improvement extends to lead times-to-change.
- The lead times for decision processes between R&D, marketing, and production should be short. This implies that cross-functional teamwork is routine.

Also in Figure 6–4, "Suggestions per employee" is certainly not the only measure that reflects the health of employee involvement. However, if the number of suggestions per employee rises well above the level of "background radiation" that it occupies in most companies, the employee involvement policies of the company must have many other healthy signs as well.

Selecting performance measures is work. Many top managements too far from the action want to specify too many of them in detail. Better to select only a few that appear to represent improvement policies very well. Let those closer to the action select the exact measures that describe improvement processes that are linked to the overall goals, but with the approval of others.

A good discipline is to have an organization's internal customers and suppliers agree on the performance measures that adequately represent their improvement policies. Then distortion in their use is less likely. Discussion focuses on policies and processes rather than on tweaking numbers, and improvements charted are more likely real. Top management's task is to be sure policies and processes are consistent.

## Oceans of Indicators

Despite effort to keep measurements simple, there is a tendency to have an ocean of performance indicators, yet not be capable of quantifying performance in an area considered important. Before becoming conscious of its importance, companies may keep no record of the length of time taken to develop a new product, nor even to have a development process specific enough that a definition of such a measure is possible. The same organization may be drowning in a sea of useless cost-accounting variances.

People cannot pay attention to each indicator in an ocean of them. They can concentrate on only a few. If an organization attempts to restrict data gathering to that which can be taken by line personnel as an adjunct activity, more thought must be given concerning which indicators are truly important. In addition, performance indicators that one posts personally tend to be more meaningful.

Even after doing this, a moderate-sized manufacturing company is apt to have hundreds or thousands of performance indicators if they count every chart at every station. Even the summaries of these will add up to a hundred or more indicators, not including accounting data. Table 6–1 is an abbreviated version of such a list of performance measures. They are classified into seven categories, but several indicators could pertain to more than one area.

## Comparing Organizations

The graphical comparison of Figure 6–4 gives better insight than the numerical comparisons of Table 6–1 even though the columns of figures are accompanied by trend indicators. The only advantage of Table 6–1 is comparison across a broader spectrum of measures, and that it shows values, although mere volume of measurements does not always add to understanding. For instance, knowledge that Division C's instruments demand more precision than the other two divisions alters interpretation of results.

While there are several reasons to make comparisons between similar operations, one of the most useful purposes is learning for mutual improvement. Benchmarking has become the popular term to describe this, and should include comparing trends in context. That is done in Table 6–1, but rather poorly. Comparing data in more detail over time, as in Figure 6–4, would be better, but the data become voluminous displayed that way. A format such as Table 6–1 is next best.

The data in Figure 6–4 and Table 6–1 are rigged to show that Division A has superior performance almost everywhere except in financial results. If a company is rated tops in quality in its market, logically it should be profitable also, but that does not necessarily follow. A company can blow a year's profit from excellent operations and customer service in one bad judgment on currency-exchange rates, for instance. And some of the best operations are in very stiff competition that holds down prices, so the primary financial beneficiary of excellence is the customer.

Although Table 6–1 contains many measurements, comparison in context is only made by following up on "why" questions provoked by the measurements—or without measurements in some instances. For example, what did Division A do to make their suggestions take off? The days of receivables seem high in all divisions, but Division B seems best. What are they doing differently? Should all divisions look elsewhere for comparisons and ideas for improvement?

There is too much data in Table 6–1 to discuss potential answers to *why*-questions for all measurements. We cannot concentrate on improvement by every measure at once, but we can describe our company performance by a balanced set of measures. One reason to make comparisons between organizations is to make improvement in one specific area. When making general comparisons, do it across all six categories of improvement goals, plus using other measures for balance.

For comparison with other organizations, data must be basically similar, but not always kept everywhere by the same rigid format. Such a practice may be self-defeating. For example, in Table 6–1, the interpretation of "repair and retest" may not have the same meaning in all three divisions.

**TABLE 6–1**
**Performance Indicators for the Three Electronic Instrument Divisions**

|  | A | B | C |
|---|---|---|---|
| Sales, $ millions | $100 N/A[d] | $200 N/A | $300 N/A |
| Production output, $ millions | $60 N/A | $100 N/A | $180 N/A |
| Percentage sales growth, last 5 yr. avg. | 34% N/A | 41% N/A | 6% N/A |

*Quality*

| | A | B | C |
|---|---|---|---|
| *Customer:* | | | |
| Percentage new units returned as defective | 0.110% F[a] | 0.240% F | 0.788% F |
| Quality ranking in industry by customer survey | First F | Second N[c] | Third N |
| Estimated mean time between failures, years | 23.4 F | 6.3 F | N/A N/A |
| Emergency service calls for instruments on first 3 years warranty, percentage | 2.6% F | 3.6% F | 12.3% N |
| *Internal:* | | | |
| Final test; repair-and-retest rate, percentage | 0.28% F | 1.24% F | 8.8% N |
| After-solder PCB reject rate, ppm | 908 F | 2493 F | 3873 F |
| Fallout: (Total parts not finishing process) / (Total parts starting process) | 3.1% F | 7.3% F | 14.3% N |
| *Suppliers:* | | | |
| Percentage of suppliers certified | 96% F | 63% F | 24% F |
| Incoming inspection reject rate, percentage | 0.04% F | 0.33% F | 1.23% F |

*Dependability*

| | A | B | C |
|---|---|---|---|
| On-time arrival rate | 93% F | 87% F | 51% U[b] |
| On-time shipping rate | 99.6% F | 98% F | 73% N |
| Delays/month for part shortages | 3 U | 9 F | 114 U |
| Schedule index (% days final production is within 10% of daily requirement) | 100% N | 95% F | N/A |

*Waste*

| | A | B | C |
|---|---|---|---|
| $ Value added/(Total headcount) | $214,000 F | $162,000 F | $138,000 N |
| $ of production/ft$^2$ | $5182 F | $5316 F | $3205 N |
| Total team projects completed | 48 F | 37 F | 26 F |

Trend of last three measurements:
[a] F = Favorable   [b] U = Unfavorable
[c] N = Not clear   [d] N/A = Not available or not found

**TABLE 6–1** *(concluded)*

|  | A | B | C |
|---|---|---|---|
| *Flexibility (Lead times)* | | | |
| Throughput time [runout calculation] | | | |
| ($ in WIP)/($ output rate) in days | 6 N | 12 F | 22 F |
| Total inventory, days on hand | 17 F | 27 F | 43 F |
| PCB lot size | 1 N | 1 F | 8 N |
| Average customer lead time, work days | | | |
| (Backlog in units)/(Daily output in units) | 43 N | 55 F | 62 N |
| Longest supplier lead time, work days | 40 F | 60 U | 65 N |
| Percentage of "Hard Parts" engineering changes | | | |
| completed in under 4 weeks | 52% F | 13% F | 5% U |
| Development time, new model, months | | | |
| (Concept meeting to first unit produced) | 11 F | 18 F | 23 N |
| *Innovation* | | | |
| Percentage products and models new within 2 years | 76% F | 47% F | 35% U |
| Percentage customers considering div. a tech ldr. | 84% F | 52% F | 12% U |
| *People* | | | |
| Absentism, all employees, percentage unscheduled days | 1.1% N | 1.3% N | 4.6% F |
| Percentage employees participating on teams: | | | |
| Production employees | 59% F | 26% F | 15% N |
| All others | 83% F | 35% F | N/A |
| Number of suggestions/ total headcount | 1.52 F | 0.12 U | 0.09 N |
| Percentage of employees trained: | | | |
| Team leadership | 100% N | 87% F | 34% F |
| Statistical process control | 94% F | 57% F | 19% F |
| Personal computer tools | 34% F | 7% F | 0% N |
| *Financial* | | | |
| Return on investment | 16% N | 17% F | 24% F |
| Debt/Equity | 32% | 47% | 67% |
| Days of receivables | 62 U | 43 U | 59 N |
| Average age of purchased equipment, years | 2.1 F | 2.9 N | 3.8 F |

Division C must select and add some of the components during final test, but the other two divisions do not. In this instance, the differences in repair-and-retest rate percentages may be less interesting than noting that Divisions A and B seem to be improving, but Division C is not.

Where an organization lacks a measurement, it probably does not regard a performance criterion as very important. For instance, under "Dependability" Division C shows an "N/A" for schedule index. That probably means that the division pays little attention to schedule execution. It could also mean that the number Division C keeps cannot be transformed to a basis similar to the other two divisions.

Because of these kinds of issues in making comparisons, the separate divisions (or even separate companies) should see mutual benefit in making comparisons in order to exercise the discipline necessary in keeping measurements translatable on a common basis.

### Strategy and Balanced Measurements

The scope of manufacturing excellence has not typically addressed broader strategic issues. Neither have most of the measurements associated with the new manufacturing. However, it increasingly appears that operating capabilities need to be competitive just to execute a business strategy effectively. An organization like Division C in Table 6–1 finds itself unable to do much more than "blow smoke" at a head-on competitor whose performance is improving as well as Division A. Operations are the fundamentals from which strategies are executed.

The data given in Table 6–1 is balanced in the sense of covering the six categories of operating improvement, but it still will not satisfy anyone interested in comparing the businesses of the divisions. For that, an analyst would want more marketing indicators, such as price comparisons, market shares, status of field-representative training, advertising approach, sales coverage, and so on. Knowing the technical changes in the business is also important. Likewise, full disclosure financial statements would be helpful.

Then there are strategic factors. Is the division backed by "deep pockets?" Do they have technical interaction with leading sources of advancement? Can they form strategic partnerships? Are there potential liability problems, or contingencies that are troubling? Will they be helped or hurt by pending legislation or regulation?

It is very easy to become engrossed in making operations improvements and forget the big, strategic picture. (In reverse, many analysts who regard themselves as astute in strategic analysis overlook the staying power possible in an organization that is extremely strong in operating fundamentals—particularly when those fundamentals extend across all functions of the company, including R&D, sales, collections, and even legal representation, as well as production.)

## Evaluation: How Do We Know What Good Is?

Comparison with recognized leaders is one way to know if performance is good, but comparison with unbalanced measurements or with a narrow focus is misleading. In the early 1980s companies often made that mistake in comparing themselves to the Japanese experience. One-dimensional comparisons led to silly conclusions. The difference was thought due exclusively to quality circles, or to *kanban* systems, or to whatever first caught the attention.

One of the better and more tested frameworks by which to evaluate performance is the basic approach used by the examiners for the Malcolm Baldrige National Quality Award. Three factors guide comprehensive evaluation of processes and performance:

1. *Approach.* In quality evaluations, approach refers to the degree in which activities are preventive and systematic. Companies create processes to prevent defects before they happen. More broadly, approach means that people doing the work are proactive in establishing customer satisfaction in almost any way it can be considered.

2. *Deployment.* Are good approaches used in every part of a company, or do indicators reflect spottiness—islands of vigorous improvement in seas of mediocrity?
3. *Results.* Outcomes are positive and appear to be sustained. Customers are satisfied in ways they are aware of and in many they are not really aware of. Evaluating results is judging performance-indicator trends and comparisons with benchmarks and ultimate goals.

Quantification of judgment is on a scale of 0–100, though a 1–10 or other scale might serve as well. Roughly, a zero means that measurements do not exist, and performance as well as the measurement of it consists of bald, unsubstantiated assertion. A 50 rating indicates the beginning of a sound system. Performance is tending toward the right direction, but may not be consistent or fully deployed. The manufacturing world is loaded with 50s.

A 100 rating means that the approach is sound, deployment complete, and results are the best. The evaluator cannot think of a way to improve. There are not many 100s.

Perhaps the best part of this evaluation framework is that it provides a systematic way to view overall performance. To use the framework, evaluators must look at trends, make comparisons, and consider the situation in context.

## Ben Franklin, Priorities, and Improvement

Ben Franklin's approach to personal improvement was to work on one virtue at a time, concentrating on it for several weeks until it became a habit. Having conquered one petty vice, such as overeating, he then began to concentrate on another, meanwhile attempting not to backslide on those already mastered.

Continuous improvement in manufacturing takes place by the Ben Franklin method. The long range objective is overall improvement of the capabilities of an organization, but people cannot concentrate on everything at once. Management must set priorities. Measurements follow.

Working down the high-frequency causes of problems as summarized on a Pareto chart is an example. For an example, an objective is to reduce throughput time, and a company has worked on setup time reduction, material movement, material signaling systems, and other obvious factors that delay flow of material through a plant. Finally they create a Pareto chart showing the reasons why machines in a plant cannot be set up to make the part wanted at the time wanted, as shown in Figure 6–5. (This is a mundane example. New product tool-

**FIGURE 6–5**
**Comparative Charts: Causes of Setup Delay in a Fabrication Plant**

1 = Necessary material not
present
2 = Tooling not ready
3 = Quality problem delays
job now on machine
4 = Delay to check first piece -
gage or instrument problem
5 = Machine maintenance
6 = Not enough workers
present to make setup
7 = Another setup takes priority
8 = Material handling equipment
not present
9 = Error made - incorrect setup
10 = Other reasons

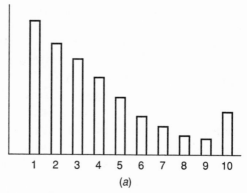

(a)

First chart plant summary pareto
chart: Causes of setup delay

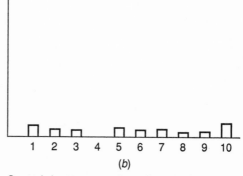

(b)

Second chart two years later: Causes of setup delay
charted in the same sequence as the first chart

ing delays, for instance, might be a much more strategically important objective.)

Then improvement effort concentrated on high-frequency reasons for delay of setup. After each month's Pareto summary was compiled, the production managers urged concentration on the top three items on the most recent chart. However, more than urging was sometimes necessary because once people became proficient diagnosing and overcoming problems causing delay of material reaching the point of use, for instance, they liked to continue working on those problems. People like to repeat success. The Pareto charts had to be posted and emphasized in order to shift attention to different problems; like the maintenance and availability of material handling equipment on short notice—after reviewing operations to be sure material handling equipment was really needed.

Finally after months, Pareto chart *A* was transformed into results shown in chart *B*. There were no more outstanding high-frequency causes of setup delay. Now what?

Perhaps working on causes of setup delay is no longer something that should receive management emphasis. Instead we should look for activities that will contribute to a permanent, preventive fix of the problems that were formerly identified as delaying setups. Several of the causes involved maintenance. Perhaps the development of preventive, predictive maintenance could become the overall priority to be reviewed with a Pareto chart.

In the meantime, how does one continue to be sure that gains related to setup delay are not program-of-the-month gains that are either superficial, or soon lost? Chart work-in-process inventory as days or hours on hand. If it begins to increase, recheck the delays of setup. Look at defect-rate data from the plant. If defect rates continue downward, setup delays should not often occur for quality reasons. But there is no reason to emphasize Pareto charts on which there is no follow-up.

Eventually, an overall improvement process will return to the setup delay, if progress in other areas continues long enough. Then real excellence is not only to correct any new, high, Pareto items, but to preclude anticipated causes for setup delay that have not even been experienced yet. However, there's no point doing that until improvement on other goals has balanced out. The use of measurements for improvement should promote this kind of process.

# INDEX